Romantic Climates

Anne Collett • Olivia Murphy
Editors

Romantic Climates

Literature and Science in an Age of Catastrophe

palgrave
macmillan

Editors
Anne Collett
English Literatures & Creative Writing
University of Wollongong
Wollongong, NSW, Australia

Olivia Murphy
Department of English
University of Sydney
Sydney, NSW, Australia

ISBN 978-3-030-16240-5 ISBN 978-3-030-16241-2 (eBook)
https://doi.org/10.1007/978-3-030-16241-2

This Palgrave Macmillan imprint is published by the registered company Springer Nature Switzerland AG
The registered company address is: Gewerbestrasse 11, 6330 Cham, Switzerland

We dedicate this book, with love and hope,
to the children who will live the challenge of the future—
Ada Ruth Murphy Wood
(born 24/08/2017)
Eloise Anne Hui-Lin Iu
(born 25/09/2018)

Acknowledgements

Much of the research included in this volume was first aired at two events held in 2016 at the University of Sydney and Victoria University of Wellington. Thanks are due to the English department and especially its chair, Liam Semler, for supporting the Sydney symposium, and to the Faculty of Law, Humanities and the Arts, University of Wollongong, for their financial assistance.

In Wellington, for hosting the '1816–2016 Year Without a Summer Bicentenary Symposium', we gratefully acknowledge Te Puna Mātauranga o Aotearoa/National Library of New Zealand. For the exhibition of 1816 related texts, we thank Anthony Tedeschi, Curator of Rare Books and Fine Printing, Alexander Turnbull Library. For their financial assistance, we are indebted to the Faculty of Humanities and Social Sciences, Victoria University of Wellington.

We want to thank everyone who attended the symposia, including the late Paul Brock, whose kind encouragement inspired our commitment to bringing this collection together.

Ben Doyle and his team at Palgrave encouraged this project from the beginning. We thank them and our anonymous reader for their considered engagement and generous criticism.

Shane Greentree's capable and indefatigable assistance made it possible to complete our revisions and prepare the final manuscript with a speed beyond anything we could have hoped, and we thank the School of Letters, Arts and Media at the University of Sydney, and in particular Umberto Ansaldo and Sarah Taylor, for their financial and administrative support to bring the project to completion.

This volume reflects some of the work of members of the Romantic Studies Association of Australasia, and we hope it stands as a testament to the strength of Australasian Romantic studies. All of us owe a debt of gratitude to RSAA's members and to its president, Heidi Thomson.

Finally, we thank our families for their love and forbearance.

Wollongong and Sydney
December 2018

Anne Collett and Olivia Murphy

Contents

1 **Romantic Climates: A Change in the Weather** 1
Olivia Murphy

2 **Domesticating Climate: Scale and the Meteorology of Luke Howard** 17
Alexis Harley

3 **Wordsworth in the Tropics of Cumbria** 33
Elias Greig

4 **Keats and the Poetics of Climate Change, 1816 and Beyond** 59
Nikki Hessell

5 **'Out of Season': The Narrative Ecology of *Persuasion*** 75
Amelia Dale

6 **'This Thing of Darkness': Reading Atmospheric Disturbance in Matthew Lewis's *Journal of a West India Proprietor*** 91
Anne Collett

7 **When the Earth Moves** 113
Clara Tuite

8 **Utopia or Dystopia? The Romantics in Switzerland, 1816** 141
Steven Hampton

9 **Metaphor and the Unprecedented: Byron's 'Darkness' and Responding to Ecological Disaster** 159
James Phillips

10 ***Orlando*'s Romantic Climate Change** 173
Thomas H. Ford

11 **Afterword: Ghosts of 1816** 191
Gillen D'Arcy Wood

Works Cited 201

Index 215

Notes on Contributors

Anne Collett is Associate Professor of English Literatures at the University of Wollongong, Australia. She has edited *Kunapipi: Journal of Postcolonial Writing and Culture* since 2000 and has written extensively on postcolonial poetry and women's writing and visual arts. Most recently, she edited (with Russell McDougall and Sue Thomas) the collection *Tracking the Literature of Tropical Weather: Typhoons, Hurricanes, and Cyclones* (Palgrave, 2017) and (with Leigh Dale) *Postcolonial Past & Present* (2018).

Amelia Dale is a lecturer in the School of Languages and the Australian Studies Centre at Shanghai University of International Business and Economics. Her monograph *The Printed Reader: Gender, Quixotism, and Textual Bodies in 18th-Century Britain* is forthcoming in June 2019 with Bucknell University Press. Her work on quixotic narratives and eighteenth-century literature and culture has appeared in *Studies in 18th-Century Culture* and in edited collections with Palgrave and Delaware University Press.

Thomas H. Ford is Lecturer in English at La Trobe University working on Romantic ecophilology and the natural history of poetry. Recent publications include *Wordsworth and the Poetics of Air: Atmospheric Romanticism in a Time of Climate Change* (2018) and *A Cultural History of Climate Change* (2016).

Elias Greig focuses his research on the intersection of literary and political representation in Britain in the 1790s. He is in the process of turning his PhD thesis, '"Retracing the Revolutions": Wordsworth's Poetry, 1793–1797', completed at the University of Sydney in 2017, into a monograph.

Steven Hampton centres his research on the multilingual and transnational nature of cultural and literary production during the eighteenth and nineteenth centuries, in particular the interrelations between literatures in English, French, and German. His focus is on the national epic as it was reconstructed, rediscovered, or invented during the Age of Revolutions in Europe.

Alexis Harley is Lecturer in English at La Trobe University. She is the author of *Autobiologies: Charles Darwin and the Natural History of the Self* (2015), editor of an anthology of nineteenth-century commentaries on William Blake and associate editor of *Life Writing*. She is co-editing a collection of essays on bees in nineteenth-century literature.

Nikki Hessell is an associate professor in the English Programme at Victoria University of Wellington. She has published widely on the intersections between Romanticism, print culture studies, and indigenous studies. Her latest book is *Romantic Literature and the Colonised World: Lessons from Indigenous Translations* (Palgrave, 2018).

Olivia Murphy is a postdoctoral research fellow in English at the University of Sydney. She is the author of *Jane Austen the Reader: The Artist as Critic* (Palgrave, 2013) and, with William McCarthy, the co-editor of *Anna Letitia Barbauld: New Perspectives* (2013). In 2015 she organized the symposium *Romantic Climates* at the University of Sydney.

James Phillips is Associate Professor of Philosophy at the University of New South Wales, Sydney. He is the author of *Heidegger's Volk: Between National Socialism and Poetry* (2005), *The Equivocation of Reason: Kleist Reading Kant* (2007), and *Sternberg and Dietrich: The Phenomenology of Spectacle* (2019) and the editor of *Cinematic Thinking: Philosophical Approaches to the New Cinema* (2008).

James Renwick is a professor in the School of Geography, Environment and Earth Sciences at Victoria University of Wellington. He has over 30 years' experience in weather and climate research, from weather forecasting at MetService to climate change research at the National Institute of Water and Atmospheric Research (NIWA) and Victoria University. His main field is large-scale climate, including the El Niño/La Niña cycle and climate impacts on New Zealand and the Antarctic.

Heidi Thomson is Professor of English Literature at Victoria University of Wellington. Her book *Coleridge and the Romantic Newspaper: The 'Morning Post' and the Road to 'Dejection'* was published with Palgrave in 2016. As the president of the Romantic Studies Association of Australasia (RSAA), she organized the interdisciplinary conference about *1816–2016: The Year Without a Summer, a Bicentenary Symposium* (Wellington, 2016).

Clara Tuite is Professor of English at the University of Melbourne. Her most recent book is *Lord Byron and Scandalous Celebrity* (2015). She is editing *Byron in Context* (for Cambridge University Press) and, with Claudia L. Johnson, is writing *30 Great Myths About Jane Austen* (for Wiley-Blackwell). She is also completing a project on literary Romanticism and the media of romantic love.

Gillen D'Arcy Wood is Langan Professor of Environmental Humanities at the University of Illinois, where he is Associate Director of the Institute for Sustainability, Energy, and Environment (iSEE), and directs the Program in Environmental Writing. His recent work, in its 'eco-historical' mode, performs cultural archaeology across spatial and temporal scales and between disciplines from literary history to the Earth and atmospheric sciences. His book *Tambora: The Eruption That Changed the World* (2014) has been widely influential in ecocriticism and climate studies and was recognized in Book of the Year awards by the *Guardian*, the *London Times*, and the American Society for Literature, Science, and the Arts. His forthcoming book, *First Ice: The Antarctic Discovery Voyages, 1838–42* (Princeton University Press), reconstructs the Victorian-era South Polar expeditions as an original encounter with a precariously glaciated Earth and climate change.

Dialogue: Weaving a Web of Awareness

Heidi Thomson writes: In June 2016 a symposium commemorating the bicentenary of the Year Without a Summer (1816) was held at the National Library in Wellington, New Zealand.[1] Earlier that year, in March, a similar symposium was held at the University of Sydney by the Australian and New Zealand Society for Eighteenth-Century Studies (*ANZSECS*). The topic of the Wellington symposium was the effect of the catastrophic Mount Tambora eruption and its literary manifestations, in particular, and Romantic climates, more generally. Our main inspiration was Gillen D'Arcy Wood's *Tambora: The Eruption That Changed the World* (2014):

> After perhaps a thousand years' dormancy, Tambora's devastating evacuation and collapse in April 1815 required only a few days. It was the concentrated energy of this event that was to have the greatest human impact. By shooting its contents into the stratosphere with such a biblical force, Tambora ensured its volcanic gases reached sufficient height to seriously disable the seasonal rhythms of the global climate system, throwing human communities worldwide into chaos. The sun-dimming stratospheric aerosols produced by Tambora's eruption in 1815 spawned the most devastating, sustained period of extreme weather seen on our planet in perhaps thousands of years.[2]

As an organizer I invited my Victoria University colleague, climatologist James Renwick, to speak about 'climate and major volcanic eruptions'. As a result of the symposium, James and I started a conversation about the disconnection between the existence of compelling data on climate change and the implementation of practices which address the problems. Our questioning revolves around this: how do we remedy the discrepancy between the incontrovertible evidence from scientific data about the destructive effects of climate change, on the one hand, and the unwillingness, or inertia, of human beings to respond constructively to well-founded predictions of inevitable catastrophe, on the other hand? The kind of small-scale historiography we normally engage in is not particularly instructive for future-oriented awareness and decision making about climate change. Yet the need for humans to plan and take action now to prevent the total exhaustion of the earth, our habitat, is more compelling than ever. Our conversation led to the idea of a short written dialogue about this issue, so, in the beginning of 2017, we exchanged the views below, largely by email or in the coffee shop of Victoria University of Wellington.

Heidi Thomson: James, you regularly go out in the field to collect data and to discuss the increasingly worrying findings with international experts. What was the purpose of the March 2017 meeting in Ushuaia in Tierra del Fuego, Argentina, sometimes referred to as the 'Fin del Mundo' or 'End of the World'? What was the experience like?

James Renwick: I travelled to Ushuaia as part of an international panel studying polar weather and climate.[3] We discussed melting glaciers, thawing permafrost, and disappearing sea ice, as well as all the technologies we have to monitor and model changes. Satellites tell us in great detail on a daily basis what is going on with temperatures, rainfalls, ice cover, clouds, winds. Weather and climate models cover the globe with fine-mesh grids, simulating our turbulent atmosphere and oceans on scales from hours to centuries. We know very precisely what is happening in our world and where it is leading. Yet I am reminded that our wonderful intellectual achievements, our rational scientific minds, have helped divorce us from the very world we live in. Once 'the environment' becomes something to be studied objectively, we place ourselves outside of that environment. We forget that this is a convenient fiction. Humanity is as much a part of the natural world as are the trees growing outside my window. We build our apartment blocks and factories just as birds build nests or beavers their dams.

Heidi Thomson: Part of the challenge seems to be that the story of the satellites, what they 'tell us in great detail on a daily basis', falls on deaf ears as far as the implementation of change is concerned. Mere data, no matter how compelling, are ignored if they are not communicated in human terms. Ironically, because of technological developments, we now have the opportunity to travel quite easily to many places which manifest the effects of climate change. Or, if we can't physically get ourselves there, we can find out about the effects through various forms of online media, where the transience and the brevity of the medium create their own problems. We like to think that we are merely visitors to sites of pristine natural beauty, but we inevitably leave more than just footprints. How did you perceive Ushuaia as a place of observation and reflection, James?

James Renwick: In Ushuaia, my hotel looked out on Beagle Channel, the stretch of water navigated nearly 200 years ago by Robert FitzRoy's *Beagle*, which also carried Charles Darwin as a passenger. Those travellers would not have seen the sprawling small city that exists here today, but the mountains to the north and south would have looked almost identical. The colonies of sea lions and cormorants peppering the rocky islands in the Channel would have looked on as indifferently in the 1830s as they do today on the tourist boats that ply these waters. Looking eye to eye with a sea lion one morning, I had the sense that these creatures are fully adapted to their surroundings; they are part of the ecosystem that surrounds them and if left undisturbed will go on living in the same way indefinitely. Conversely, my presence on the tourist boat was a marvel of technology involving thousands of kilometres of air travel, electronic communication, commerce, boat building, and navigation. I am the anomaly here. Today I saw on one of the islands the remains of a Yámana dwelling built around 4000 years ago by the people who populated these shores through most of the Holocene period. One must assume they lived a subsistence lifestyle, with very different concepts of growth and change to ours and very different technology. Yet they are in a sense my ancestors and, given access to weapons and fuels and building materials, they would likely have grown to dominate their local environment, as we do. Today, we tend to survive in spite of our surroundings rather than because of them. Wherever we go we insulate ourselves from our surroundings with sturdy well-heated houses, with cars and ships and aeroplanes. We almost forget that our technological society is as embedded in the natural environment as is the colony of cormorants on the rocky islet in Beagle Channel. In our state of hubristic denial, we are constantly surprised by the power of nature to wreak havoc in our communities. We forget how vast are the energies that drive our weather and climate, despite our sophisticated ability to quantify the manifestations of nature's power.

Heidi Thomson: I like your reference to the 'vastness' of the energies. Scale and size have always been a source of fascination for artists and writers, particularly when the notions of geological time and evolution started to become part of scientific, philosophical, and artistic discourses. Those discourses used to be much more closely connected under the umbrella of 'natural philosophy'. Percy Shelley, writing 'Mont Blanc' in 1816, only thirty years after the first ascent of that mountain in 1786, speculates: 'Is this the scene / Where the old Earthquake-dæmon taught her young / Ruin? Were these their toys? or did a sea / Of fire, envelope once this silent snow? / None can reply—all seems eternal now'.[4] As ice caps and glaciers visibly diminish from year to year, there is no semblance of eternity anymore. What do you make of the accelerated accumulation of 'unprecedented' weather events, James?

James Renwick: Exactly. As we heard in Ushuaia, glaciers that have supplied vital water to communities for centuries are now disappearing. After thousands of years of slumber, the oceans are again rising, and climates are changing around the globe. These are events humanity is not used to dealing with, and we do not appreciate just how rapidly extreme events can change with even seemingly small shifts in average climate. Half a metre of sea level rise makes the one-in-hundred-year coastal flood occur every few months. Two degrees of warming make previously unknown extreme heat waves commonplace. Adjusting our collective thinking to these new realities may be one of the most difficult tasks in coming to terms with and dealing with a changing climate.

Heidi Thomson: For hundreds of years, we have been ensconced in a mindset which equates productivity with profit and progress with access to unlimited, supposedly inexhaustible resources. While the Romantic period is still primarily associated with an artistic interest in the development of the individual Self, it was also the era of the industrial revolution. The Romantic era—the 'Age of Wonder' as Richard Holmes terms it—saw an exponential growth in the need for fossil-based energy, as transport and machinery developed on an industrial scale. What made the Romantic period different from ours, however, was the more relaxed, interdisciplinary flow of ideas between arts, literature, and science. The various branches of knowledge, admittedly the prerogative of a small elite, were less narrowly specialized and more a matter of public interest. Public demonstrations and lectures of scientific discoveries drew enormous crowds and inspired the paintings of Joseph Wright of Derby and the novels of Mary Shelley. But even then, the largely unquestioned norm of 'productivity' and 'progress' relied on a confidence in an inexhaustible pool of 'natural resources'. When poets like William Wordsworth realized that the 'world is too much with us', that 'getting and spending, we lay waste our powers', they may have been articulating their concerns from an anthropocentric notion of Self, but they surmised, as Byron did in *Darkness*, that it was 'not all a dream', that perhaps not all was right with the idea of progress.[5]

Since then the rift between the arts, humanities, and sciences has grown immeasurably through the adoption of largely mutually exclusive taxonomies, specializations, and categorizations. The lack of resonance between the arts, humanities, and sciences makes the joint effort to combat pure greed very difficult. For those who master the language of science, it provides a constant stream of revelations and wonders. For those who are not attuned to scientific discourses, that same stream seems impenetrable and meaningless. While the idea of so-called interdisciplinary projects is the academic, and grant-getting, flavour of the decade, specialization in increasingly narrow areas is the norm for most fields of study. Identifying disturbing data and making them somehow comprehensible to a wide audience, with a view to a solution, is a massive challenge for the humanities and the sciences. Our inertia is grounded, ironically, in our conservatism, in a resistance to change, and in a desire to read catastrophic events as anomalies, even when those anomalies are occurring at ever smaller intervals. Instead of thinking of our joint cooperation as interdisciplinary, we believe that we need to think of our intellectual and social projects as interdiscursive in which we graft the practices and language of one field onto another in order to enhance interactive and mutually understandable communication.

Percy Shelley compares the skylark in his poem to a visionary, intellectual poet who manages to persuade his audience of truths they had not even imagined, bringing the world to 'sympathy with hopes and fears it heeded not'.[6] What is the challenge for twenty-first century prophets? I know that you maintain an active Twitter feed, James, and I wonder what your thoughts are about the various types of media available to us now? What works, in your opinion? And what doesn't work?

James Renwick: Anything that connects with people is what works, in my opinion. Today, much of our mass communication is through social media, blog posts, and ever-revolving news feeds, and being a part of that world is an important component of communicating science today.[7] However, as fake news or alternative facts circulate the globe around the clock, the task of communicating factual information, let alone knowledge or wisdom, becomes ever harder. Of course, a Twitter post hardly changes the world! Such things can help to raise awareness, to give a 'heads-up' on new scientific observations and understandings. A social media presence must be balanced by a physical engagement, in communities and sectors of society, through public presentations and conference participation, through being available to people. This is where we have the opportunity to truly connect with others, to hear and to share knowledge and stories that shed light on what concerns us. Story-telling is really where all communication begins and ends.

Heidi Thomson: A rhetoric of persuasion is often connected with a convincing, palatable story. Within the arts and the humanities, a story can take many forms: theatre, performance arts, novels, poems, paintings, music can all contribute to a better recognition of our place in the world. While scientific data paint a picture of sorts, they do not connect the dots for our flawed human understanding. But the successful integration and transformation of the data into novels, poems, plays, paintings, and music may produce a shift in awareness, a push towards deeper reflection and sensibility, a prompt for change. This is not so much a matter of 'sugaring the pill' as of transforming and uniting the fabric of our knowledge, weaving a web of awareness which engages the mind and the heart beyond our monomaniacal interests.

James Renwick: Stories, and emotional connection, are how we communicate. That is why I am interested in engaging with the story-tellers: the writers, artists, musicians, the 'humanities'. Turning scientific knowledge into poetry or sculpture helps us all understand and to see a path to the future. The 2016 'Pacific Climate Change Conference' at Victoria University of Wellington was a great success precisely because it brought together all those streams of thought and expression in a space where stories could be shared.[8] The aim of our next conference, in February 2018, is to repeat and amplify the experience, to meld scientific data and understanding with human emotion and expression, with daily realities.[9] Seeing the human stories behind 'the data' is, to me, the way to bridge the gap between the sciences and the arts.

Heidi Thomson: Yes, I agree. The integration of pressing environmental, scientific, and social concerns into works of art runs the risk of turning art into propaganda, but ground-breaking art usually manages to escape the trap of didactic prescription. While so-called objectivity, or objective distance, is considered a virtue in scientific research methodology, the arts may help us realize that we are far more implicated in any process of observation or experimentation than we dare to admit. When a group of privileged English travellers ended up stuck in the Villa Diodati on the shore of Lake Geneva, they dispelled boredom with a story-telling competition. An eighteen-year-old conceived of a gothic tale in which a creature is fashioned out of bits from the graveyard and the slaughterhouse. It is shocked into life with electricity by its maker, who subsequently abandons its creation, to the perdition of many. Mary Shelley's *Frankenstein* is still read as a tale cautioning us to take responsibility for what we set into motion, just because we can. As the essays in this collection prove on a number of counts, we ignore those stories at our own peril. As 'unacknowledged legislators', poets, like scientists, need to convince the world of solutions to 'hopes and fears it heeded not'.[10]

Wellington, July 2017

James Renwick
Heidi Thomson

Notes

1. *1816–2016 'The Year Without a Summer': A Bicentenary Symposium* was held on Friday, 3 June 2016, at the National Library of New Zealand. The event was organized by the Romantic Studies Association of Australasia (RSAA) and Victoria University of Wellington, in collaboration with the Alexander Turnbull Library.
2. Gillen D'Arcy Wood, *Tambora: The Eruption that Changed the World* (Princeton and Oxford: Princeton University Press, 2014), 8.
3. The meeting of the Executive Council Panel of Experts on Polar and High-mountain Observations, Research and Services (EC-PHORS), Ushuaia, Argentina, 21–24 March 2017.
4. Percy Bysshe Shelley, 'Mont Blanc. Lines Written in the Vale of Chamouni' (ll.71–75), in *Shelley's Poetry and Prose*, ed. Donald H. Reiman and Sharon B. Powers (New York: Norton, 1977), 91.
5. 'The world is too much with us; late and soon' (ll. 1, 2) in William Wordsworth, *Poems, in Two Volumes, and Other Poems, 1800–1807*, ed. Jared Curtis (Ithaca: Cornell University Press, 1983), 150; Byron, *Darkness* (l.1) in *The Poetical Works of Lord Byron* (London: Oxford University Press, 1957), 95.
6. 'To a Sky-Lark' (l.40) in Shelley, *Shelley's Poetry and Prose*, 227.
7. James Renwick tweets about climate change @CubaRaglanGuy.

8. For the outcome statement of the 2016 'Pacific Climate Change Conference', see http://www.victoria.ac.nz/vicpasifika/our-community/events/climate-conference/outcome-statement-pacific-climate-change-conference.pdf.
9. See http://www.confer.co.nz/pcc2018/.
10. 'A Defence of Poetry' in Shelley, *Shelley's Poetry and Prose*, 508; 'To a Sky-Lark' (l.40), 227.

CHAPTER 1

Romantic Climates: A Change in the Weather

Olivia Murphy

In April 1815 the eruption of Mount Tambora on the island of Sumbawa, in the Indonesian archipelago, triggered a global climate disaster. In Southeast Asia more than 100,000 people were killed in the explosion and the massive tsunami that followed. The effects of the eruption were to be felt far beyond its devastated epicentre. Clouds of ash filled the sky, obscuring the sun and causing rapid temporary global cooling. Across the world crops failed, livestock died in the fields and natural cycles were thrown into chaos. The disastrous consequences of the eruption would unfold over several years, causing widespread confusion, suffering and death. For Europe, still recovering after decades of war and their bloody culmination at Waterloo, the dark skies and frigid temperatures of 1816 seemed to portend a troubling future. The effects of this environmental calamity can be seen in the cultural productions of the period, in some of the most significant works of the movement we call Romantic.

More than two centuries later, the 'Year Without a Summer' offers us a starting point from which to reconsider both how the Romantics responded to the changing climates of their day and to think about how these climatic events shaped the development of Romanticism itself. Climate, be it political, literary or meteorological, is what we live *in*, or live *through*; it is whatever we take for granted. Only when the atmospheric

O. Murphy (✉)
Department of English, University of Sydney, Sydney, NSW, Australia
e-mail: olivia.murphy@sydney.edu.au

A. Collett, O. Murphy (eds.), *Romantic Climates*,
https://doi.org/10.1007/978-3-030-16241-2_1

becomes the catastrophic—as it has so obviously in our own times—do we allow it space at the centre of our culture and of our thinking. The Year Without a Summer brings the disruptive possibilities of climate changes into sharp focus, but even more powerful, if less spectacular, forces of environmental transformation had been at work long before 1815. Regardless of volcanic activity at the equator, in Western Europe anthropogenic climate change had been accelerating in the Romantic period as a consequence of the Industrial Revolution. Contemporary debates around the identification of an 'Anthropocene'—a new geological age marked by human environmental damage—often take 1800 as its starting point.[1] As Tobias Menely points out, however, the 'epochal switchover to fossil energy was substantively underway in the eighteenth century', with the 'amplifying feedback between coal, capital investment, and technological development' traceable to the Restoration period.[2] The impact of the unprecedented transformation of the carbon cycle caused by the burning of fossil fuels from the Industrial Revolution through to today has yet to be fully felt, although we are coming to think of mass extinctions, climatological turmoil and rising sea levels as unavoidable. In the Romantic period, such changes were more subtle still. The sudden catastrophe of Tambora's eruption makes legible those slower processes which would come, in our own day, to have such devastating effects. Mass industrialization and deforestation, exponential increases in the human and livestock populations and ever more widespread imperial ventures that globalized the practices of Western European industrialized capitalism were all features of the Romantic world, and their consequences are still unfolding. The Romantics themselves understood at least part of what was at stake in the mutually supportive ideologies of empire and marketplace. Kate Rigby identifies a 'pervasive Romantic resistance' to the discursive frameworks that justified the 'treatment of the earth and its 'natural resources' as freely available to be appropriated, traded and made-over by merchants and manufacturers', where the 'liberty of human property owners'—including, we would add, owners of enslaved human property—'is to be enlarged at the expense of the colonization, commodification and exploitation of those (ever expanding) portions of nature they claimed as theirs'.[3]

Like the Romantics, we too live in a time of rapidly changing climate, and like them we are preoccupied by fears of extinction, by questions of how our species does and should interact with our environment and by the devastating strangeness of our unpredictable world. Like them, we face a political sphere changing at an unprecedented rate, where the pieties and absolutes of our youth can be overturned in an instant. Like them, we must face the reality that far-distant events can affect our daily lives in unexpected ways and that our own actions may have unintended, far-reaching consequences. By understanding the impact of climate on Romanticism, we seek also to understand how we might, as students of literature and critics of culture, intervene productively in the challenging climates of our own time.

As the events of 1815 and its aftermath make clear, climate is a planetary phenomenon with myriad local consequences. The world-wide impact of Tambora's eruption demonstrates the ways in which Romanticism must be located not in a particular corner of Europe, but as a global movement disseminated through networks of trade, empire and exploration. This collection grew out of events held in two cities—Sydney, Australia, and Wellington, New Zealand—that have been shaped by Romantic-era colonialism. The lands of the Gadigal people of the Eora nation, and that of Te Ūpoko o te Ika a Māui on New Zealand's North Island, have now survived the impact of this colonialism for more than two centuries and face new challenges today as the climate of the southern hemisphere becomes ever more unpredictable.

This volume seeks to uncover how our own thinking about climate and catastrophe has been formed by the ideas of Romantic writers and thinkers and how these same ideas might once more be harnessed to assist us in the new climate challenges facing us in the present. As Timothy Morton notes in the introduction to his *Ecology Without Nature* (2007), 'the literature of the Romantic period, commonly seen as crucially about nature ... still influences the ways in which the ecological imaginary works'.[4] We cannot help but think Romantically when we think about Nature. We are as apt as Jane Austen's Anne Elliot to 'fall into' Romantic quotation when confronted with seasons of mist, lonely clouds and seas, whether sunless or silent. The conceit of 'Nature' is itself a troubling Romantic fallacy, especially insofar as it is contrasted with civilization, or seen as fundamentally distinct and distant from human life. Personified as a feminine 'mother', Nature partakes of patriarchal assumptions that it is, or ought to be, passively nurturing: any activity unwelcome to humans is interpreted as a threatening aberrance from the 'natural' norm (imagined as 'the contagion of a mother's hate' in Shelley's *Prometheus Unbound*).[5] It is these ideas, as inherently self-contradictory and nonsensical as scrutiny reveals them to be, that nevertheless impinge upon our capacity to understand earth's ecosystem in the twenty-first century. As Morton points out, however, the source of much of what is problematic or confused in our thinking about our world can also be the source of challenges to such ossified thinking. Romantic-era texts, he writes, 'not only … exemplify, but also … do *not* accord with the various syndromes and symptoms that emerge from this very period. At the precise moment at which the trajectories of modern ecology were appearing, other pathways became possible'.[6] The chapters in this volume do some of the work of recovering those alternative pathways, as well as exposing where Romantic thinking has led us astray.

The impact on Romantic thinking of climate catastrophes in general, and of the 1815 eruption of Mount Tambora in particular, has been raised to international attention by Gillen D'Arcy Wood's *Tambora: The Eruption that Changed the World* (2014), which has led to other investigations of its aftermath, such as Robert Markley's research into the 1816 British embassy to the Chinese Emperor.[7] For decades, however, Romanticism has been a fruitful field for ecocritical approaches to literature, such as the 'Green Romanticism' established by Jonathan Bate's *Romantic Ecology* (1991). The chapters in this volume follow more recent work by Morton, Menely, Clark, Louise Westling, John Parham and others that seeks to ground criticism of late eighteenth- and early nineteenth-century cultural production in an understanding of both Romantic-era environmental realities and intellectual cultures. Heidi Scott takes a similar approach in *Chaos and Cosmos: Literary Roots of Modern Ecology in the British Nineteenth Century* (2014), as do the contributors to *Romantic Sustainability* and *Romantic Ecocriticism* (both 2016), in order to consider what we might learn from Romanticism and its response to climate.[8]

Ironically, despite its volcanic origins far off in the southern hemisphere, the Year Without a Summer is now most closely associated with the work of a small group of friends who gathered in 1816 at the Villa Diodati on the shore of Lake Geneva. The friends included Lord Byron, Byron's doctor John Polidori, Percy Shelley, Mary Wollstonecraft Godwin (who would marry Shelley after his first wife's suicide later that year) and Godwin's stepsister Claire Clairmont, at the time pregnant with Byron's child. Many of the group were in self-imposed exile from England—Byron because of the notoriety of his recent divorce, Shelley and Godwin because of their illicit relationship—but they were also part of the first wave of British tourists to explore Europe after the end of the Napoleonic Wars. The Diodati party was a famously productive one: Byron wrote the poem 'Darkness' and the third canto of *Childe Harold*, and a fragment of a ghost story by

Byron was reworked by Polidori into *The Vampyre* (1819), the first vampire novel. The most famous work produced that summer was by the 18-year-old Mary Godwin, whose *Frankenstein* (1818) would be published under her married name.

Frankenstein's success has largely overshadowed Mary Shelley's other works and come to dominate our understanding of the Year Without a Summer, the gloomy, gothic world of post-Tambora Europe. It has come to stand in the popular imagination as a cautionary tale about the limits of science and the drive to subjugate natural processes, such as death, to human desires. More recently the novel has come to be read in light of pressing questions of who, or what, counts as 'human' and of how to be human in a hostile environment. Many of the chapters in this volume consider these issues from perspectives other than Shelley's, showing that such questions were part of the atmosphere of 1816.

This is not, however, the only story to tell about the world of 1816 and the impact of climate on Romanticism.

As the contributions to this volume demonstrate, climate is an inescapable aspect of Romantic writing and thinking. Ideologies and experiences of climate inform everything from scientific writing to lyric poetry and novels. Crucially, such generic distinctions do not necessarily hold for a time before the institution of today's disciplinary boundaries, and Romantic literature cannot be read in isolation from the scientific developments of the period. The then-emerging disciplines of geology, botany, meteorology, comparative anatomy and palaeontology have their bearing on novels and poetry, and vice versa. An understanding of climate is inseparable from an understanding of Romanticism, and yet we are still learning to turn our attention to climate phenomena in Romantic writing, phenomena that have been long overlooked because of their very ubiquity. Chapters in this collection consider the works of Jane Austen, John Keats and William Wordsworth, along with the Diodati circle and less well-known figures, such as the scientist Luke Howard. The chapters in this volume offer new readings of Romantic literature, and literary Romanticism, that seek to assess the ways in which such texts negotiate questions of climate.

Today we are used to thinking about climate as, fundamentally, a problem for the sciences and for disciplines such as law and political science that we hope might hold out answers to what is coming to be seen as humanity's greatest challenge. The chapters in this volume speak to the ways in which the humanities, and especially literary studies, have their own part to play in understanding and responding to climate. One of the greatest strengths of literature, and thus of literary studies as a discipline responsive to and responsible for interpreting literature, is its ability to draw on and catalyse ideas, ideologies and metaphors from a seemingly infinite range of human experience and expertise. Literature makes it possible to imagine conversations across disciplines, across geographic and political divides and across the centuries. In the dialogue that opens this volume, Heidi Thomson and James Renwick argue for a new, 'interdiscursive' approach to considering climate, an approach which draws on the capacious and promiscuous methodologies of literature to bring together insights from a range of intellectual disciplines to meet the unprecedented challenge of climate change. The scientific distancing of the environment as a set of topics to be studied, Renwick argues, has been a boon to measurement and data-gathering, but it has yet to offer feasible solutions to the great problem of how to keep the planet habitable for the plants and animals that currently call it home. Thomson and Renwick argue that the scientific mindset that has dominated our thinking since the Romantic period itself can be enlarged and enhanced by thinking of the environment not just as a series of phenomena to be observed but also as an infinite web of experiences to be lived. This is the hallmark of Romantic writing about that most contested term, 'Nature'. It may also prove to be Romanticism's most valuable legacy. Through re-learning the capacity of literature to imaginatively situate us—as individuals, as a species—in relation to our world, we might finally become capable of abandoning the self-destructive anthropocentric worldview that reduces the world to its component parts, its exchange value in global capitalism. 'Getting and spending', wrote Wordsworth, 'we lay waste our powers;/Little we see in Nature that is ours'. Wordsworth believed that the 'world'—the world of society and economics—'is too much with us', but it is equally true that the world on which we live and depend is not sufficiently 'with' us, and we are insufficiently 'with' the world.[9] By bringing Romantic literature into the conversation about climate, we hope that this volume will model the kind of interdiscursivity that Renwick and Thomson point to as the last best hope for our species and its place on the planet.

Part of the effort required for such interdiscursivity to work is that of rethinking what we 'know' about Romanticism and about a world in which the effects of the Industrial Revolution were only beginning to be felt. Reimagining ourselves into the lived experience of the Year Without a Summer requires considerable effort, not least because our understanding of climate benefits from some two centuries of scientific progress, and can now be measured, tracked and even predicted with increasing confidence. As Alexis Harley's exploration of the work of Luke Howard shows, Romantic-era meteorologists and weather watchers were only beginning to create the tools for measuring climate data that we now take for granted. Harley considers Howard's work on clouds: his attempts to distinguish and define different cloud categories. This work, both appreciated and mocked in his own time, reflects the epistemological requirement of the sciences to 'fix' a particular object or phenomenon—even one so obviously unfixed as clouds—in order to render it a viable subject for scientific investigation. Howard's work, as Harley shows, brought clouds, the subject of poets and painters, into the realm of scientific scrutiny, continuing the Enlightenment project of interpreting the natural world via scientific rather than religious or artistic means.

Howard's efforts to similarly define, categorize and measure climate, principally through recording temperature variations, continued this work of subjecting the natural world to scientific rigour. As Harley argues, however, the very act of measurement can lead to disturbing conclusions, as an effort to demonstrate the predictability of the seasons—to determine reliable points of comparison—was frustrated by the capriciousness of the weather itself. Howard's measurements of London's weather are an invaluable resource for researchers but, as Harley shows, they reveal as much about the ideologies surrounding Romantic-era climate science as they do about the climate itself. Howard's attempt to take baseline measurements of weather in the metropolis came up against the coldest winter ever recorded. In his observations, we can see the confidence of Enlightenment science give way to the radical uncertainty of highly variable data. Howard's stubborn insistence on material accuracy and producing an unsensational record of climate patterns was challenged by the very sensationalism of the weather itself, his attempts at combatting empirical mysteriousness floundering in the face of sublime extremes of climate.

The empirical once again challenges the metaphysical in Elias Greig's consideration of Wordsworth's Lake District, where Greig reveals the threat inherent in the landscape to be as material and historical as it is allegorical. The ongoing dangers of the Lake District, as starkly represented by the deaths and disappearances of modern-day tourists seeking to enjoy the landscape Wordsworth made famous, force us to acknowledge this cold, unfeeling environment—an environment that kills as well as sustains—as the contextual basis of Wordsworth's poetry. Taking critics such as Aldous Huxley, Alan Liu and Jonathan Bate to task, Greig attempts to rupture the conventional understanding of Wordsworth's 'nature' as chiefly moral, metaphysical and theological, thus exposing as a critical complacency those readings of Wordsworth that choose to ignore the 'stark and often stubborn materiality' of the poet's depictions of nature. Greig draws on the poetry of the 1790s to argue for a 'complex and disturbing engagement with the physical universe', one that is sympathetic to the plight of the poor and the dispossessed, for whom the landscape offers only insufficient support along with the ever-present threat of destruction. He sees Wordsworth's early work—poems including 'Salisbury Plain', 'The Ruined Cottage' and 'The Borderers'—as establishing a relationship between, on one hand, the amoral agency of nature and its ruinous effects, and on the other the immorality of a political climate in Britain that funded ongoing war with France by imposing taxes on a population already impoverished by bad harvests and depressed trade. The 'ruined cottage' is thus a metonym for a nation in material, historic decline. Greig's reading undercuts a critical mythology of Romanticism, calling for a reinterpretation of 'Tintern Abbey' and the 'Immortality Ode' in light of Wordsworth's radical materialism, taking into account the cold indifference of the natural world's impact on individuals, and the implicit criticism of a Tory unconcern for the welfare of the English populace. In his reading of Wordsworth's poetry of the 1790s, Greig uncovers the harsh climate of immoral social neglect that leaves the poor exposed to the amoral workings of Nature, a 'careless' nature more recognizably akin to that found in the poetry of Tennyson or even Ted Hughes, than one that we might think of as conventionally Romantic.

The ways in which Romantic poetry mediates the natural world (as distinct from mimetically representing it) are brought into sharp focus by a comparison of Wordsworth's version of nature with that found in Keats's poetry. Investigating Keats's productive poetic years from 1816 to 1819, Nikki Hessell considers the value he derived from the literal lack of poetic material in the 'backward Spring' of 1816. Driving the poetic imagination to fill that material lack, Hessell argues, Keats produces a poetry that finally blooms in 'the delightful weather' of 1819, in which the maturing poet was able to reflect on the benefits of his earlier uncertainty. Rather than reading tentativeness in the poems of 1816 as stemming from a class-based lack of poetic experience and perspective, Hessell considers the material pressures on Keats's poetry in that year. The impact of climate change on familiar flora and fauna produced its own radical uncertainties, and Keats transmutes the historic absence of botanical specimens—in the cold post-Tambora weather, many plants relied upon by medical science simply did not grow—into 'metaphorical usefulness' in which the poetic and imaginative faculties are stimulated to replace what is missing. This botany of absence generates a rhapsodic poetry that imagines more than is there, turning in upon itself to reflect on the art of poetry-making. Through her close reading of Keats's work in these years, Hessell thus disrupts our commonplace understanding of Romanticism and its relationship to Nature.

The complex ways in which Romantic writers challenge conventional thinking about the relationship between art and nature, illuminating the failures and possibilities of both, are also at play in Amelia Dale's essay, which shows how one famously empirical writer grappled with the unstable climates of the nineteenth century's second decade. In her reading of Jane Austen's *Persuasion* (1818), Dale shows how Austen's nuanced use of weather in her last completed and most explicitly historically grounded novel becomes a way of thinking through time, memory and indeterminacy. The irregularities of the climate are reflected in *Persuasion*'s secondary characters, seeming to age too rapidly or not at all, and in its heroine Anne Elliot, who in the course of the novel experiences a 'second spring' of youthfulness and beauty. The novel offers complex meditations on the slippage between the apparent stasis and continuity of a conservative society and the disruption and dynamism of unstoppable historical currents. Set in 1814, before Tambora's eruption and Napoleon's escape from Elba, but written in the wake of Waterloo and the miserable summer of 1816 (a summer that would prove to be Austen's last), *Persuasion* asks of its readers a kind of chiasmus of memory. In order to sympathize with its characters, we must enter into their certainties—that war is followed by peace, and winter by spring—yet the knowledge that these certainties have been disturbed by extraordinary global events, both political and climactic, troubles our interpretation of every confidently expressed prediction the characters have to offer.

Dale shows how Austen's novel forces both her characters and her readers to move backwards and forwards in time, its close engagement with seasonal transitions promising not dependable continuity but rather radical instability. The novel, argues Dale, folds in on itself in ways that challenge our idea of time as linear. *Persuasion* offers its readers a superficially static society in which, conversely, nothing stays the same. It also creates an environment in which places, texts and even moments can be revisited again and again, the past and future collapsing in on the present, all complicated by the novel's own historicism. Dale's ecocritically informed reading of *Persuasion* demonstrates the power of the novel to allow its readers to think in similarly complex ways about the interrelationship of the present, past and future, with implications for our understanding of our place in time and the world. The novel's fantasy of second chances and quiet revolutions might serve as a model for responding to the proliferating challenges we now face.

The world of *Persuasion*, with its casual discussion of the West Indies, Bermuda and the Cape (of Good Hope), opens up the new world of far-flung empire, a world crossed and re-crossed with unprecedented ease by new technologies of navigation and one which would repeatedly challenge both the values and eventually the very nature of England as 'home'. In 1816, when Austen was writing *Persuasion*, the gothic novelist Matthew (Monk) Lewis left the increasingly gothic party at the Villa Diodati to undertake the first of two visits to the slave-run Jamaican sugar plantations he had inherited on his father's death. Anne Collett shows how Lewis's *Journal of a West India Proprietor* records his emotional and moral turbulence during a period of unsettled political and meteorological weather, from late 1815 through to his death from yellow fever in mid-1818. The posthumous publication of the *Journal* in 1834 coincides with the year in which the Slavery Abolition Act took effect throughout the British Empire, initiating a new phase of imperial expansion and racialized control. Lewis's *Journal* reflects some of the difficulties—what Collett hears as an intellectual 'jangling'—faced by a slave owner with sincerely held abolitionist principles, one who sought to improve conditions for the slaves on his plantations, without considering emancipation as a feasible possibility. While Lewis reiterates myths of the cannibalistic, literally blood-thirsty slaves with a relish reminiscent of *The Monk*'s most outrageous passages, his preoccupation, and even at times identification, with those—like the ship's pig—destined to be eaten, reveals his persistent unease with profiting from the mortification of human flesh and the misery of human suffering.

In the final entry of his journal, Lewis notes the perturbation that follows his final command as an *in situ* West India Proprietor, 'the whole air' of the island being 'rent with noises of all kinds and creatures'. Collett examines the ways in which Lewis's account is imaginatively and, in some senses, ideologically indebted to Caliban's island in *The Tempest*, a focus for imperial anxieties from early modern times. Lewis recalls Shakespeare's island that is 'full of noises', but whereas the noises in *The Tempest* take the form of 'sweet airs, that give delight and hurt not', the noises of colonial Jamaica are disquieting and disturbing. While Lewis seems to turn to Prospero as a model of magnanimous, princely authority and command of 'his' island and its enslaved inhabitants, the magician's robes are an uncomfortable fit. As Collett argues, the Elizabethan voyages of discovery, long classed as 'heroic', and the associated colonization of the West Indies and its accompanying slave trade, are damned by a Romantic revolution insisting on the moral authority of liberty and equality. Lewis finds himself in an unsupportable intellectual position, one he was unable to remedy. His discomfort registers in the journal through the riot of noise, colour and smell of the West Indies that assaults his senses and his sensibilities. His own half-mocking empathetic attempts are rendered supremely ironic by events. While Lewis may joke about the dismalness of his fate, he assumes his superiority to the trafficked people and animals who have shared his voyage. Beginning (anticipating Joseph Conrad) in the 'black melancholy' of the Thames, like so many others Lewis ends his journey in the limbo of the middle passage. His body, buried at sea, becomes just another victim of the slave trade.

Lewis's *Journal* demonstrates the surprising connections—across oceans, across species and across history—being made by the Romantics in the summer of 1816. Clara Tuite considers various kinds of 'teleconnections'—'deferred or distant relations between cause and effect'—that influence literary composition. Taking into account Byron's *Manfred*, and the 'storm of scandal' that accompanied the poet in his exile from England, Tuite investigates the entwined emotional and geological climate of the Year Without a Summer, in its writer's engagement with discourses of catastrophe. For Tuite, *Manfred* itself is volcanic, an example of 'the literary work as natural catastrophe or revelation'. Catastrophe and revelation were two of the most pressing scientific concerns of 1816, as discoveries in geology and what we would now call palaeontology raised new, and at the time unanswerable questions about the history and fate of the earth and its inhabitants. Tuite shows how Byron draws on the catastrophism theories of scientists such as the comparative anatomist Georges Cuvier, firmly siding against the 'gradualists' who believed that the world changed only slowly over massive spans of time. While Darwin's adoption of the gradualist worldview would ensure its prominence in the nineteenth and into the twentieth century, more recently uncovered evidence of the earth's history points to sporadic, devastating events of destruction and mass extinction, such as the one we are living through today.

The 'mysterious agency' of *Manfred* speaks to what was for the Romantics the inexplicability of such dramatic change, but Tuite argues that it is also a way of thinking about teleconnection—in this instance, as a means by which the exiled Byron-in-Geneva was able to impact, through his writing, events in an England to which he would never return. His anxious attempts to send geological samples from the Alps back to his half-sister and infant daughter in England also speak to this attempt to transverse geography, to cross time and space through sheer feeling. As Tuite writes, Byron's characterization of love in his writing from this period also partakes of the rhetoric of catastrophism: love can bring together people separated by great distance, like a fold in the crust of the earth; it can also 'wreck' everything around it, like an earthquake or a volcanic eruption. Like the new discoveries of Cuvier and his colleagues, love can also bring the past back to life in surprising and disruptive ways, making a nonsense of our complacent assumptions about continuity and gradual progress. Tuite brings Austen's *Persuasion* into her reading here, to consider the ways in which the novel, like *Manfred*, 'both thematizes and experiments with time', with love operating as a catalyst for disrupting linear time. Austen's hero and heroine overcome the pain of the past and the restrictions on gender relations in their society in order to come to a blissful understanding that, until the end of the novel, seems impossibly out of reach. The 'magical thinking' that brings together Austen and Byron—who never met, but whose works echo one another's in surprising ways—shows how this kind of imaginative play can foster interdiscursiveness between individuals separated by history.

History also brings individuals together, however, and it is one assembly in particular that has come to stand in our imagination for the Romantic response to the Year Without a Summer. As Stephen Hampton's discussion of the literary output of the famous group gathered at the Villa Diodati in 1816 shows, interrogating critical commonplaces and reinvestigating well-known territory with a sense of the particular and the historical can bring nuance to our understanding of Romantic thinking. Hampton shows how works such as the diaries kept by Byron and his friend John Cam Hobhouse, and Mary Shelley's novel *The Last Man* (1826), all reflect on the status of the Switzerland the friends first experienced in that dark, wet summer. Shelley's better-known novel *Frankenstein* (1818) began, so the tale has it, in a ghost story competition between Byron, Hobhouse and the Shelleys. Its complex relationship to the Year Without a Summer has now become part of the *Frankenstein* legend. Shelley's later novel *The Last Man*, however, published a decade after that momentous summer, has received comparatively little critical attention.

The Last Man envisions a future in which even the millenarianist dreams of visionaries like William Godwin and Percy Bysshe Shelley have come to pass, a world which, like post-Napoleonic Switzerland, enjoys a degree of democracy, peace and plenty unknown to Romantic-era Britain. This seemingly idyllic world, however, is progressively destroyed by an unstoppable plague that lays waste to humankind, depopulating the earth and destroying all human civilization. In contrast to the equally apocalyptic 'Darkness', Byron's poem of 1816 in which not only life on earth but all light, and finally even motion, ceases, in Shelley's imagined future it is only human beings who are destroyed. The world—with all its birds and animals, and the beautiful, even sublime landscapes of Western Europe—flourishes in the absence of humanity. Hampton reexamines these texts, along with others by the Diodati circle that make use of the potential of Switzerland as a setting, to reflect on the complexities of meaning embedded in the Swiss landscape. To the Romantics, Hampton argues, Switzerland (with its enlightened, enfranchised and newly liberated citizenry) was seen as a progressive and idealized image of an ideal future state. Underlying this image of the alpine country, however, was an older association of the mountainous region with death, disaster and even the devil himself. It is these satanic possibilities of the mountains that Byron draws on in *Manfred*, the 'dark' Switzerland of an older worldview, one brought back to consciousness by the darkness of the summer of 1816.

The mountains of Switzerland have come to seem like the inevitable backdrop to the sublime Romanticism of Shelley and Byron, one that, as Hampton points out, benefited from their rich legendary associations. The nihilism of Byron's 'Darkness' thus grows out of an encounter with place while also responding to a crisis in time. In James Phillips's reading of Byron's poem, 1816's season of backwardness gives rise to a consideration of what was missing: of the lack, 'lessness', decline and vanishing, that generate a poetic 'crisis', a crisis to which Byron responded with mocking melodrama. Where Hessell's essay on Keats is concerned with the material lack that a poet might fill, Phillips writes of a metaphysical lack that the poet despairs of filling. The conscious failure of Byron's poetry to overcome this slippage between what is ideally desired and what is materially possible forces us to question, in Phillips's words, 'the vocation of poetry and the business of interpretation' in the face of environmental catastrophe. The poet-narrator's isolated, if not impossible, perspective, from which he observes the end of the world in absolute darkness, at first glance seems to endorse the most extreme ideas of the egotistical sublime. The poet's refusal to maintain a relationship between confessional and descriptive statement, however, leaves the descriptive material without a champion, sustaining and even widening the gulf between the totality of all that is, and all that can be imagined, and the limits of what can be encompassed in poetry.

Phillips makes the claim that 'Darkness' is a poem of mocking despair, not so much for the end of the world, as for the incapacity of the poet to connect the individual self to the world that self inhabits, but from which the conscious self is always somehow separate. Phillips reads 'Darkness' as a poem that at once recognizes and performs the ultimate incapacity of the Romantic poet to transgress the bonds of individual subjectivity. It is thus less a poem about the end of the physical world, and more a helpless articulation of the shortcomings of language to describe what has never been described, and what turns out to be indescribable. Phillips sees in 'Darkness' an attempt to bridge the gulf between the poem and the world which is the poem's ostensible subject, an attempt that flounders in the face of the wreck of the poetic ego. The failure of clarity and perspective intimated by Byron's experience of the 'stupid mists—fogs—rains—and perpetual density' of 1816 anticipates the wreckage of the poetic 'I' in the imagined end of the world. 'Darkness' thus encapsulates an apocalypse that can never be, and yet somehow already is, 'a dream, which was not all a dream'. It is the dream of complete poetic engagement with nature turned nightmare, of nature itself returning to an impossible primordial nothingness in which it is incompatible with life. This despairing prediction is only possible in the realm of art, but its fulfilment in reality seems ever more terrifyingly plausible.

To move impossibly backwards and forwards through time, to reinhabit the lives of the long dead and imagine the world of the future, is one of the greatest possibilities afforded by literature. Literature is all too frequently well ahead of criticism in its ability to imaginatively reconnect with the past and consider its place in the present. Scholarly thinking about climate and Romanticism, for instance, is of relatively recent date. As Thomas Ford shows in his reading of *Orlando* (1928), however, Virginia Woolf anticipates this preoccupation by almost a century. Ford argues that in Woolf's novel, references to the weather work simultaneously on multiple levels: they are meant to be read literally, as indications of a material reality, but they also operate as metaphors—Ruskin's 'pathetic fallacy'—which Woolf deploys in order to reinvigorate the climatic metaphors of Romantic lyric poetry. Ford demonstrates the ways in which Woolf takes on Ruskin, revisiting and reimagining his ideas and deploying them with a poetic dexterity he could not have envisaged. She makes climate an agile metaphor for the end of the eighteenth-century Enlightenment; the fog and darkness of the Industrial Revolution become both the distinguishing marks and the constitutive basis of the nineteenth century and its culture. A 'damp' and a 'chill' that is at once literal and figurative settles over Britain, in contrast to the icy brilliance of the little-ice-age Elizabethan England in which the novel opens. Woolf, argues Ford, reknits the relationships between weather and culture, between material sign and imaginative signifier. In so doing she both mocks Ruskinian strictures and points out the shortcomings of a materialist realism that cannot account for climate—the inescapable, if diffuse atmosphere in which all culture must operate. Ford considers *Orlando* as a kind of reawakening of Romanticism, one that is both extraordinarily conscious of historical context and at the same time curiously free from it. Woolf revivifies what had become desiccated Romantic metaphors to serve Modernism and the novel rather than the lyric poetry which had become ossified and canonical. In doing so she models the kind of work through which Romanticism—or at least Woolf's flexible, transmutable Romanticism—can continue to contribute to our understanding of our world.

Writing a few years before Woolf, Marcel Proust had claimed that a 'change in the weather is sufficient to create the world and oneself anew'.[10] We are beginning to learn that, at least as far as the world is concerned, this is literally true. In offering new readings of Romantic literature, and literary Romanticism, that seek to assess the ways in which such texts negotiate questions of climate, it is hoped that the chapters in this volume might contribute to the work of changing ourselves, and our world, as much as we humans have already changed the weather.

Notes

1. See, for instance, Timothy Clark, *Ecocriticism on the Edge: The Anthropocene as a Threshold Concept* (London: Bloomsbury, 2015), 1.
2. Tobias Menely, 'Late Holocene Poetics: Genre and Geohistory in *Beachy Head*', *European Romantic Review*, 28:3 (2017): 307–314, 307–318.

3. Kate Rigby, 'Romanticism and Ecocriticism', in *The Oxford Handbook of Ecocriticism*, ed. Greg Garrard (Oxford: Oxford University Press, 2014), 60–84, 64.
4. Timothy Morton, *Ecology Without Nature: Rethinking Environmental Aesthetics* (Cambridge, MA: Harvard University Press, 2007), 1.
5. Percy Bysshe Shelley, '*Prometheus Unbound* I. 178', in *Shelley's Poetry and Prose*, ed. Donald H. Reiman and Neil Freistat (New York: Norton, 2002), 214.
6. Morton, *Ecology Without Nature*, 7.
7. Robert Markley, 'The Amherst Embassy in the Shadow of Tambora: Climate and Culture, 1816', in *Writing China: Essays on the Amherst Embassy (1816) and Sino-British Cultural Relations*, ed. Peter J. Kitson and Robert Markley (Woodbridge: Boydell and Brewer, 2016), 83–104.
8. Heidi Scott, *Chaos and Cosmos: Literary Roots of Modern Ecology in the British Nineteenth Century* (University Park: Pennsylvania State University Press, 2014); Ben P. Robertson (ed.), *Romantic Sustainability: Endurance and the Natural World* (Lanham, MD: Lexington Books, 2016); Dewey W. Hall (ed.), *Romantic Ecocriticism: Origins and Legacies* (Lanham, MD: Lexington Books, 2016).
9. William Wordsworth, 'The world is too much with us' (composed c.1802–1804), ll.2–3, 1, in *William Wordsworth: The Major Works*, ed. Stephen Gill (Oxford: Oxford University Press, 2000), 270.
10. Marcel Proust, 'Le Côté de Guermantes' (1922), trans. C.K. Scott Moncrieff as *The Guermantes Way*, pt II (London: Chatto and Windus, 1922), 49.

CHAPTER 2

Domesticating Climate: Scale and the Meteorology of Luke Howard

Alexis Harley

One February evening in Tottenham Green, just north of London, Luke Howard laid a wet finger on the iron railing outside his home. His flesh instantly congealed into the metal. His quicksilver thermometer, suspended a little above the snow, stood, as Howard put it 'at *Zero*'. By 11 pm, a spirit thermometer in the same position indicated that the temperature had dropped to −4 °F. Early the next morning, Howard read the lowest minimum temperature London had seen in 20 years: −5 °F (−20.6 °C).[1] That night of the 9–10 February 1816, after four consecutive days of temperatures well below 0 °C, London's inhabitants were plunged into a cold so profound it has not once reoccurred in the subsequent two centuries.[2] As Howard wrote down the lowest minimum of the coldest year[3] of one of the coldest decades ever recorded across the northern hemisphere,[4] he had no way of knowing that Tottenham Green was enmeshed in a global cooling, that this cooling was owed primarily to a massive volcanic eruption the previous year, and that it would disarray the normal patterns of Spring and Summer weather to disastrous effect. He did not yet know that in Western Europe it would produce what Gillen D'Arcy Wood has described as 'biblical-style inundation' during the Summer growing season,[5] wreaking havoc with the lives of animals and

A. Harley (✉)
Department of English, La Trobe University, Bundoora, VIC, Australia
e-mail: a.harley@latrobe.edu.au

A. Collett, O. Murphy (eds.), *Romantic Climates*,
https://doi.org/10.1007/978-3-030-16241-2_2

people. He did not know that across England crops would fail, livestock would be slaughtered for want of silage, and throughout Europe many people, poorly supported by governments that had spent their reserves in the Napoleonic Wars, would starve to death.[6]

But even though Howard had no way of knowing at the time that he was recording the early effects of an accelerated, worldwide, if ultimately short-lived, climatic shift, it is nonetheless remarkable how unremarkable he found this extraordinary temperature. Howard had been making daily rounds of his meteorological instruments, slate and pencil in hand, since 1807. There was no precedent in his decade of daily records for this stunningly cold morning. And yet, he wrote in his notes at the end of the month, he had been 'prepared to expect' it, by the low temperature of the previous evening and the clearness of the sky. By noon, the temperature had reached 25 °F, still well below freezing, and Howard observed that it was 'pleasant, by contrast' and that 'it was easy to keep warm in walking without an upper coat'. In fact, he claimed, even 'at *Zero* ... the first impression of the air on the skin was not disagreeable'.[7] In the second volume of his *Climate of London* (1818–1820), published four years after the event, he was obliged by a discussion of London's temperature range to reflect again on the coldest day in his records. In the intervening period, Howard had gained considerably more knowledge about what that cold Winter's day had portended. In August and September 1816, Howard had taken a nine-week tour from Amsterdam to Geneva (during which period his London observations were conducted by a friend). Writing up the months' London observations, he added details from his travels:

> [F]rom the sources of the Rhine among the Alps, to its embouchure in the German ocean, and through a space twice or thrice as broad from east to west, the whole season presented a series of storms and inundations. Not meadows and villages alone, but portions of cities and large towns, lay long under water: dikes were broken, bridges blown up, the crops spoiled or carried off by torrents, and the vintage ruined by the want of sun to bring out and ripen the fruit.[8]

Perhaps on account of this further experience, the anodyne description written in 1816 of a 'not disagreeable' atmosphere becomes a little more charged, in 1820, in one of many images throughout Howard's work where his meteorological terminology coincides with an aesthetic lexicon. 'So cold was the surface on the 9th at noon', he recalls, 'that a bright sun, contrary to its usual effect in our climate, produced not the least moisture in the snow, the polished plates of which retaining their form, refracted the rays with all the brilliancy of dew drops'. But even here Howard represses the significance of the event with a matter-of-fact haste. Respecting 'the *intense cold* of the 9–10th of Second month 1816', Howard notes, 'I need not enlarge here having given already a pretty long note on the subject, under Table 115, vol 1'.[9]

This apophatic sentence—effectively an enlarging on his intention not to enlarge upon the coldest day of his records—betrays Howard's embarrassment in representing the weather of 1816. The year was full of outlier events: months that were rainier than any other month in his records, temperatures that distorted Howard's averages, frosts that occurred in the wrong seasons, and minima that bent his ranges out of shape. *The Climate of London* was nothing without its data. Almost the entirety of the first volume and a good proportion of the second consist of monthly tables of daily measurements, under which appear a statistical synopsis for each month and a set of explanatory or interpretive notes. Howard could not exclude the data of 1816, and yet these data disrupted the work's stated purposes: to furnish meteorology with 'general principles', to delimit the variations of the climate of London within fixed periodic cycles, and to make predictable its vacillations.[10] The weather of 1816 also interfered with what seems to have been the larger purpose behind Howard's work, to counter the Romantic representation of the physical world as resistant to human reckoning, as immeasurable, unstable, and unknowable—not so much because humans are poor observers (Howard acknowledged that they often were), but because of its own volatile agency.

In his *Romantic Rocks, Aesthetic Geology* (2004), Noah Heringman argues that it was the same set of cultural practices in the period between 1770 and 1820 that shaped both a poetry stocked with awe-inspiring rocks and the emerging science of geology. Changing attitudes 'towards the earth's material and toward materiality itself' resulted in 'not only a body of poetry obsessed with mountains, but also a geology steeped in aesthetics'.[11] The sublime rhetoric associated with mountains still infuses geological writing as late as 1830, when Charles Lyell published the first volume of *Principles of Geology* (1830–1833). Howard was to meteorology what the uniformitarian geologists, Lyell, and before him, James Hutton, had been to their discipline. Both Lyell and Hutton sought to demonstrate that geological phenomena had been formed through the action of consistent processes and that the future effects of these ongoing processes could be anticipated. The alternative geological view, in which geological phenomena were supposed to be the result of erratic convulsions, ejaculations, and floods, Lyell called 'catastrophism'. Howard might have described himself as a uniformitarian meteorologist, but, unlike Lyell, he seems to have preserved very little of the sublime poetics with which the literary culture around him was treating weather. Seth Reno has claimed the extensive influence of Howard's cloud taxonomy on Romantic culture: William Wordsworth and Percy Shelley had both read Howard's essay *On the Modifications of Clouds* (1803), and possibly the painter John Constable was steered towards clouds by the popularity of Howard's work.[12] But Howard does not seem to have been concomitantly influenced or inspired by Romantic literary culture. I suggest that this is because, unlike the geologists whom Heringman studies, Howard did not participate in the changing attitudes to 'materiality itself' that were shaping both literary and scientific culture. Rather than embracing the deconstruction of the mind-matter binary that manifested in the discourse of the sublime, Howard took a trenchantly mechanistic view of nature.

In *The Climate of London*, he explicitly takes up arms against the idea of agential nature, complaining about 'a mode of expression', used by many 'men of science', in which '*Nature*, personified, is made to do every thing, while the Great Author of nature is never mentioned or alluded to'. Indeed, Howard continues, 'Surely no well informed mind can now imagine that the chain of causes and effects, which we contemplate in Natural philosophy, could ever arrange and move itself'.[13] As Peter Hanns Reill argues, the mechanistic view of nature associated with the Enlightenment had given way in the second half of the eighteenth century to what Reill calls 'Enlightenment Vitalism'. Where the mechanists had entrenched the mind-body binary, the vitalists 'sought to bridge or dissolve this dichotomy by positing the existence in living matter of active and self-activating forces, which had a teleological character'.[14]

The 'men of science' who embraced vitalism, according to Reill, included such prominent natural historians as Georges Cuvier and Alexander von Humboldt and the chemist, Antoine Lavoisier. But 'men of science' did not have a monopoly on the representation of nature as active and self-actualising; the figuration is also anchored in Romantic literary culture. And the sublime climatic events of the Summer of 1816 seem to have spawned a fresh cohort of vitalist writings. The obvious example is Mary Shelley's depiction in *Frankenstein* (1818), the novel she began writing in Switzerland that dismal Summer, of a life force that animates the hitherto dead flesh of Frankenstein's creature.[15] Mary Shelley's partner, Percy Bysshe Shelley, saw Mont Blanc only once during their six-week tour of the Swiss Alps, because it was so immured in cloud.[16] But it nonetheless gave rise to 'Mont Blanc: Lines Written in the Vale of Chamouni' (1817/July 1816), one of many contemporary works that unpick the mind-body binary and depict agential matter. 'The ever-rolling universe of things' acts on or with the mind rather than being subject to it.

Mobilising the discourse of the sublime, Shelley's poem suggests that matter actively resists rationalisation or measurement. The mode of the sublime, as Edmund Burke describes it in his *Philosophical Enquiry into the Origin of Our Ideas of the Sublime and Beautiful* (1757), entails the baffling of human reason through scalar difference: the viewer experiences magnitude, the infinite, or the infinitesimal, in ways that displace his capacity to make sense of what he is seeing. Rather than taking the measure of the ravine of Arve, the speaker apostrophises it—'Dizzy Ravine!', he cries—and thereby performs his own dizziness, as the epithet slips from the speaker's overwhelmed mind to the plunging waters of the ravine.[17] For the self-declared atheist, Percy Shelley, sublime immeasurable self-actualising nature functions in this poem as a perfectly adequate alterative to the 'Great Author of nature'. In Shelley, atheism, vitalism (or, arguably, a variant belief in agential matter), and an anti-rationalism that appear in the discourse of the sublime make for a plausible alliance. It would be too far a simplification to suggest that such an alliance was routinely pitted against an equally neat marriage of theism, a mechanistic view of matter, and rationalism. Vitalists could be, and often were, believers in a god; atheism and rationalism have enjoyed a long association. And yet this is a reasonably neat characterisation of Howard's relationship to a culture that increasingly celebrated either or both anti-rationalism and the idea of nature as an actor. Until the publication in 1844 of Robert Chambers' *Vestiges of the Natural History of Creation*, he was one of many natural philosophers who would invoke the mechanical world and God the Author as evidence for each other's existence.

At any rate, for Howard, the measurability of the world and the potential knowability of climate were not to be thwarted by the dynamics of the sublime: climate's immensity and scale are not to entail its incalculability. Labouring over the instruments at Tottenham Green, Howard strives to know and fix a climate that should be knowable (and already fixed) because it is divinely ordained. The ultimate purpose of his meteorological record-keeping, Howard noted in an essay of 1807 and again in the second volume of *The Climate of London*, was that 'we should discover a chain of causes and effects, demonstrative like the rest of creation, of the infinite wisdom and goodness of its Author'. He hoped, he wrote, that Meteorology would:

> be rescued from empirical mysteriousness, and the reproach of perpetual uncertainty; and will contribute its share to the support of a proposition, so well illustrated by some of the brightest names in science, that the 'Almighty hand, that made the world of matter without form, hath ordered all things in measure and number and weight'. (Wisd. xi. 17, 20)

For Howard, then, the measurability of the world, which follows from its orderliness, indicates the benevolence of the Almighty Hand, much in the same way that for William Paley, writing just a couple of years before, the happy mechanical arrangement of the human frame, the fitness of the eyes of birds to their purpose in spying out food, and the regularity of planetary orbits all evidenced the omnipotence, omniscience, and goodness of the deity.[18] As Howard put it, 'amidst perpetual fluctuations' the Creator has 'so adapted the means to the end, that … the balance of the great machine is preserved, and its parts still move in harmony'.[19] His attempt in *The Climate of London* to figure climate as measurable, and to know its fluctuations within the schemata of 'the great machine', serves his belief in the divine authorship of the world *and* his conviction that the nature of this divinely authored world is such that, through careful measurement and analysis, humans can obtain predictive knowledge of its workings. Crucially, human knowledge of the climate is not to supplant, but to enable people to profit from, the Great Author's benevolent ordering of it. '[I]t be a very just remark, that the seasons would not only not go on better for our purposes, but would be in utter confusion, had mankind the *ordering* of them', he writes, but 'it does not thence follow that, could we calculate their periods and foresee their extremes, both our personal safety and comfort, and the success of our labours, might not be essentially promoted by such foreknowledge'.[20]

The tangle of double negatives and the forlorn conditional mood of this sentence might owe to the fact that Howard's faith in his own capacity to calculate the periodic variation of London's climate and foresee its extremes had been drastically undermined by the peculiar climatic events of 1816. If so, they are just a couple of the many contortions, rhetorical and logical, necessitated by the tension between Howard's commitment to recording the weather through 1816, however irregular it was, and his argument for the ordained regularity and periodicity of its variation. What he observes that year is fundamentally in conflict with the theoretical claims he seeks to evidence, namely, that the climate of London is unchanging—or rather, that its vacillations are periodic—and that as the science progresses, London's climate will become accordingly predictable. His nothing-to-see-here rhetorical manoeuvres around the profound cold of 9–10 February are apiece with similar manipulations of content, statistics, and form throughout *The Climate of London*. These attempts to smooth over the significant disruption to the record are just as prone to be undone by the particularities of the day-to-day data that he records. For the data is nonetheless there, to the tune of 150 months' worth of tables. Indeed, for Wood, Howard's *Climate of London* is *the* primary source for the weather in London during the period. As he puts it, 'That the myriad legends of the "Year without a Summer" have some statistical basis in the temperature record is owed—at least in England—to a man with a strong claim of "father of meteorology"'—Luke Howard.[21]

Most climate historians acknowledge that the 'Year Without a Summer' produced significant disruption to the apparent pattern of a climate. But significance depends upon the relative scales of differences and those differentiating them, a fact that Howard exploits in the strategic averaging of his data and in his graphs, where the disproportionate axes can shrink a difference out of significance. In fact, of course, Howard's records testify constantly to difference or disruption. No two Januaries, for instance, have the same median temperature; no two Junes the same rainfall. Howard's attempt to fix the climate of London was always, 1816 or no, going to require his dismissal of certain differences as insignificant in order to produce meaningful categorical claims. In this, *The Climate of London* continues the work of his far more famous publication, *On the Modifications of Clouds* (1803), an essay that organised clouds into seven distinct categories (cirrus, cumulus, stratus, nimbus, cirro-cumulus, cumulo-stratus, cirro-stratus) and sought to pinpoint the meteorological causes and, importantly for weather-forecasters, the effects of each of these forms. Howard's labour in *The Climate of London* to chasten the 'vague and unphilosophical language' of weather watchers[22] is a continuation of his work classifying clouds. Like all classification projects, and like Howard's attempts to produce categorical generalisations about London's climate, his taxonomising of clouds involves the scalar manipulation of difference, with some degrees of difference treated as salient and others dismissed.

Unconvinced by the legitimacy of this manipulation of difference, an early reviewer of *On the Modifications of Clouds* wrote in 1805 that there is 'something of ludicrous importance in classifying and defining, with Linnéan formality, and in choice Latin, the tribe of clouds, of which the varieties are only seven'. He concluded: 'In the author's philosophy, we do not find much that calls for remark'.[23] The reviewer's concern that the 'varieties are only seven' presumably emerges from some intimation that Howard has flattened into insignificance an infinity of differences. These repressed differences bubble up in Howard's struggle to describe clearly each category he seeks to fix. We see this through the fact that even while he disputes the idea that clouds are 'ever varying and therefore not to be defined', he moves into metaphor, and his metaphors refuse to stand still. In describing what he calls the Cirrus cloud, he writes that 'They are first indicated by a few threads pencilled, as it were, on the sky. These increase in length, and new ones are in time added to them. Often the first-formed threads serve as stems to numerous branches, which in their turn give rise to others'.[24] By turns, threading, pencilling, branching, and stemming, the cloud seems to be a mixed metaphor before the fact, an anticipation of the semantic cloudiness that its inherent dynamism and resistance to categorisation will produce. This is despite the fact that Howard announces at the beginning of his essay that part of his purpose, in naming the cloud modifications, is to avoid the cloudiness of the vernacular weather-watching lexicon, riddled as it is with metaphor, and to rectify the fact that hitherto meteorological data has existed 'in a confused mass of simple aphorisms'.[25] That his descriptions of clouds remain masses of metaphor demonstrates that even as Howard attempts to create a metalanguage with which to objectify clouds, that language falls prey to clouds' refusal to hold still. Even the word Howard uses for his cloud categories, 'Modifications', troubles his attempt to make stable objects of clouds. The word is a noun, so syntactically it reifies the categories, but it has a semantically verbal aspect (the -fication suffix being used for nouns of action) as if allowing for the ongoing activity and mobility of the categories and their constituents.

On the Modifications of Clouds seeks to transform dynamic processes into stable objects. Howard argues that clouds are the legible version of their own invisible causation, and he proposes a clear meteorological language to overcome the semantic cloudiness of existing folk terminology. Unsurprisingly, perhaps, his essay does not escape the influence of a literary culture deeply invested in the weather as a cluster of tropes that signal the limits of rationalism—and deeply invested in the cloud, in particular, as a figure for mutability and indeterminacy. There is, for instance, the 'flattering cloud' that 'appears to show/The fancied semblance of a distant sail,/Then melts away' in Charlotte Smith's Sonnet 43 (1784)[26]; the Cloud in William Blake's *The Book of Thel* (1789) that 'in one hour ... fade[s] away' but nonetheless provides water for horses and dew for flowers, and is, as an object, fleeting, but as part of an intra-acting system, eternal[27]; and, in the decades following publication of Howard's essay, the clouds as figures for transience in John Keats' poems—the 'weeping cloud' that falls '[s]udden from heaven' like a melancholy fit or the 'barred clouds' that 'bloom the soft-dying day' in 'To Autumn' (1819),[28] and the multiform, highly mutable cloud of Percy Shelley's poem, 'The Cloud' (1820), which Seth Reno argues depicts 'the interdependence of all things in the natural world'.[29] I would like to suggest, however, that it is clouds themselves, rather than just the way clouds figure in literature, that disrupt Howard's attempt to objectify them. The limitations to clouds' legibility play out in his descriptions, slippery with similes, as the language he hopes will signify the cloud itself disperses and modifies. Howard's attempt to taxonomise clouds, to provide for them a nomenclature, and to fix their signification, seemingly works against everything that Romantic literary culture knows of clouds: their mobility, their transformability, their transience, and their semantic multivalency. In fact, he acknowledges and dismisses this understanding of clouds:

> If Clouds were the mere result of the condensation of Vapour in the masses of atmosphere which they occupy, if their variations were produced by the movements of the atmosphere alone, then indeed might the study of them be deemed an useless pursuit of shadows, an attempt to describe forms which, being the sport of winds, must be ever varying, and therefore not to be defined.[30]

But, he writes, despite 'the erroneous admission of this opinion' that has stood in the way of people's attention to clouds, it is not the case. 'They are', Howard says, 'subject to certain distinct modifications, produced by the general causes which effect all the variations of the Atmosphere: they are commonly as good visible indications of the operation of these causes as is the countenance of the state of a person's mind or body'.[31] Tellingly, in order even to write of how clouds are *in reality* 'subject to distinct modifications', he resorts to yet another simile.

In 1820, Johann Goethe dedicated a poem to Howard, which includes the lines:

> Er aber, Howard, giebt mit zeinem Sinn
> Uns neuer Lehre herrlichsten Gewinn:
> Was sich nicht halten, nicht erreichen lässt,
> Er fast es an, er halt zuerst es fest.
> Bestimmt das Unbestimmte.[32]

In English, Howard gives us with his new theory the most superb prize: what cannot be stopped, what cannot be reached, he touches and he is the first to hold tight. He determines the indeterminate. The line 'Bestimmt das Unbestimmte'—he determines the indeterminate, or he gives precision to the imprecise—repeats the paradox of 'Was sich nicht halten ... er halt zuerst es fest' (What cannot be held [or stopped] ... he is the first to hold tight). It is clear from Goethe's (admiring) poem that he reads Howard as imposing a fixity on something essentially unfixed, making a conceptual object of phenomena that are not, in their processual material reality, an object. In this, he is undertaking what later becomes a scientific commonplace: the practice of stabilising an object for a community of researchers to think with. Modern scientists, Lorraine Daston explains, regularly 'furnish the universe with objects that are amenable to sustained and probing investigation but that rarely correspond to the objects of everyday perception'.[33] It is interesting to note that until the eighteenth century, according to Theodore Porter, the 'objects' associated with objectivity were usually 'objects of consciousness rather than physical things; real entities existing outside of us were called subjects'.[34] Howard's meticulous assembling of daily meteorological readings, of temperature, air pressure, wind direction, rainfall, and humidity, across 150 months, bespeaks his commitment to the observation of physical things, and yet ultimately, for him, physical things, or thingings, are subordinated to objects of consciousness.

Against the culturally prevalent understanding of clouds as moving, as confused, or as polysemic, Howard's focus is on the cloud as 'an aggregate of minute drops',[35] an aggregate held together by the pressure of the atmosphere on aqueous vapour and the inertia of the aqueous vapour's particles, which resist diffusion. This is cloud as assemblage or mass or collective noun. Cloud, indeed, as cumulus or accumulation. Understood in this way, the cloud does not suggest itself as a figure for indeterminacy or transience, but for aggregation, for bringing together like components and defining them against unlike—for doing the taxonomising work that Howard had learnt from the Linnaean method in botany. The star of Howard's essay, the cloud, suggests itself as a metaphor for two competing ways of understanding nature, weather, climate, and matter. The matter of weather can be seen as separable into (at least temporarily) discrete categories, clouds as aggregates, and/or it can be seen as always in transit, always multivalent, infinitely complex in its relations. Employing one of these understandings of clouds and de-emphasising the other, Howard indicates that his scientific commitment is to the production of generalisations, to theory making, rather than to the inevitably difference-riddled data arising from an inherently mobile source. He says as much in the preface to *The Climate of London*. The purpose of his records is to amass '*data* ... to found a theory that might guide us to more certain conclusions'.[36] What to do when the data confound the very theories they are supposed to found?

'The result of my experience', Howard would write at the end of a decade in which England saw its most volatile climatic shifts in centuries, is:

> on the whole, unfavourable to the opinion of a permanent change having taken place of later time, either for the better or the worse, in the Climate of this country. Our recollection of the weather, even at the distance of a few years, being very imperfect, we are apt to suppose that the seasons are not what they formerly were; while in fact, they are only going through a series of changes, such as we may have heretofore already witnessed, and forgotten. That the shorter periods of annual variation in the mean temperature, depth of rain, and other phenomena of the year, which will be found exhibited in this volume, may be component parts of a larger cycle is, however, very possible. Otherwise, considering that the changes consequent on the clearing of woods, culture and drainage, with some other less obvious effects of an increased population, have probably by this time contributed their utmost to its improvement, I should venture to suppose, that our Climate is likely to remain for ages what it now is; and further that, in its great leading features, it differs little from what it was, when the present elevation of these islands above the sea was first established.[37]

The overt message here is that the climate of London is fundamentally stable. However, the claim of stability is hedged at every turn by remarks about temporal scale. No *permanent* change has taken place, not only *of later time* but even, probably, since the long distant first establishment of the British islands (a remark, incidentally, which, alluding to James

Hutton's account of the formation of the British isles,[38] confirms Howard's philosophical sympathy with uniformitarian geology). He likewise anticipates that the climate will remain stable *for ages*. While, on a smaller temporal scale, there has indeed been climatic variation (which Howard dismisses as 'the shorter periods of annual variation … and other phenomena of the year'), inspection across a longer range, as he will try strenuously to show, will put this variation in due proportion. Those who might think that the climate has changed, based on the observations of recent years, are constrained by their own temporality: the duration of fallible human memory forbids them from recognising that apparently unprecedented seasons are likely revisitations of previous very similar ones.

In the preface to the first volume of *The Climate of London*, Howard congratulates his readers on 'the milder uniformity and tameness of our climate'.[39] Anyone who had experienced the effect of dire corn shortages, flooding, or the almost complete annihilation of fruit crops over the previous two years would surely have cavilled at this claim of the region's climatic 'tameness'. And yet Howard manages to pepper the following notes with observations to the effect that London is 'so situated, that even in the coldest season of the year, the medium of the twenty-four hours, upon a long average of years, does not fall below the freezing point'.[40] His own data reveal that in the Winters of 1816 and 1817, not only did daily minima frequently fall below freezing point, but so did the day's medium on numerous occasions. There is no mathematical discrepancy between, on the one hand, Howard's claim about the above-freezing medium across a long-range average of seasons and, on the other, the fact that London endured a profound freeze in the Winter of 1816; but there is a discrepancy in what is implied about the providential mechanism of London's divinely authored climate. Shifting the temporal scale on which Howard conducts his analysis transforms the meaning of his data, producing the welcome illusion of a permanently equable climate, and allows him to ignore the six contiguous days in February 1816 where even the maximum did not rise above freezing point.

Foregrounding means, mediums, and averages, and repressing extremes, does much of the work needed to smooth over the disruptive climate data of 1816, but where the forbidding weather of that year cannot be completely dismissed, Howard has tried to explain it as part of a periodical variation. The second volume contains a long study of the relationship between lunar cycles and the increase and decrease of the mean temperature, focussing on the warmest and the coldest years of his study:

1807 and 1816. 'The Mean Temperature *of the year* is found to vary in different years, to the extent of full *four and a half degrees*', he admits, but, he goes on to say:

> this variation is periodical. The extent of the periods, for want of a sufficient number of years of accurate observations, cannot at present be fully determined; but they have the appearance of being completed in *seventeen* years. We may consider one of these cycles, as commencing either with 1790 or 1800, and ending with 1806 or with 1816.[41]

The conjectural relationship between the lunar cycle and the climate of London amplifies the scale of Howard's mechanistic understanding of climate. His domestic daily readings—literally domestic, for they are conducted at his home—become implicated in a cosmic mechanism, the scale of which makes petty such climatic glitches as a year without a Summer.

Notes

1. Luke Howard, *The Climate of London: Deduced from Meteorological Observations, Made at Different Places in the Neighbourhood of the Metropolis*, vol. 1 (London: W. Phillips, 1818), T. CXV. Unlike the second volume of *The Climate of London*, published in 1820, the first volume, which tabulates the weather from 1807 to 1816, contains no page numbers beyond the 38 pages of its preface and front matter. At the end of the second volume, material from the first volume is indexed alongside the monthly table of meteorological observations under which it is organised, rather than by page number: hence T. CXV for text appearing in the first volume following the 115th monthly table.
2. Scott Whitehead, who has been keeping personal meteorological records since the late 1980s, has not been able to find evidence of a London minimum less than or equal to −20.6 °C since 1816, which compares astonishingly with the coldest minimum from his own records (−10.3 °C on 12 January 1987). See Scott Whitehead, '1816: The Year Without a Summer in London', *Wanstead Meteo: meteorology-based musings about east London and beyond* (blog), November 12, 2015. https://wansteadmeteo.wordpress.com/2015/11/12/1816-the-year-without-a-summer-in-london/.
3. According to Howard's records, the median temperature for Tottenham Green in 1816 was 38 °F. Between 1807 and 1815, the median had not dipped below 47.4 °F. See Howard, *The Climate of London: Deduced from Meteorological Observations, Made at Different Places in the Neighbourhood of the Metropolis*, vol. 2 (London: 1820), 109.

4. See Michael R. Rampino, Stephen Self and Richard B. Stothers, in 'Volcanic Winters', *Annual Review of Earth and Planetary Sciences*, 16 (1988): 83.
5. Gillen D'Arcy Wood, *Tambora: The Eruption that Changed the World* (Princeton, NJ: Princeton University Press, 2014), 49.
6. For an account of the widespread food shortages across Europe, see William W. Klingaman and Nicholas P. Klingaman, *The Year Without Summer: 1816 and the Volcano That Darkened the World and Changed History* (New York: St Martins, 2013), 169–193.
7. Howard, *The Climate of London*, vol. 1, T. CXV.
8. Ibid., T. CXXII.
9. Howard, *The Climate of London*, vol. 2, 113.
10. Howard, *The Climate of London*, vol. 1, xxxvi.
11. Noah Heringman, *Romantic Rocks, Aesthetic Geology* (Ithaca and London: Cornell University Press, 2004), xiii, xiv.
12. Seth T. Reno, 'Romantic Clouds: Climate, Affect, Hyperobjects', in *Romantic Sustainability: Endurance and the Natural World, 1790–1830*, ed. Ben P. Robertson (Lanham, MD: Lexington Books, 2016), 38, 40.
13. Howard, *Climate of London*, vol. 2, p. vi.
14. Peter Hanns Reill, *Vitalizing Nature in the Enlightenment* (Berkeley: University of California Press, 2005), 7.
15. Marilyn Butler claims that while Victor Frankenstein is indeed a vitalist, Mary Shelley belittles his philosophy, representing him as a 'blundering experimenter', and is herself, at best, sceptical towards vitalism. But D. L. Macdonald and Kathleen Scherf argue that given that Frankenstein succeeds in his experiment, Shelley can be seen as (at least) open to vitalist ideas. Butler quoted in Macdonald and Scherf, 'Introduction', to Mary Shelley, *Frankenstein; or, The Modern Prometheus: The 1818 version*, 2nd edn. (Peterborough, Ontario: Broadview Press, 2005), 20.
16. Donald H. Reiman, Neil Fraistat, Nora Crook, *The Complete Poetry of Percy Bysshe Shelley*, vol. 3 (Baltimore: The Johns Hopkins University Press, 2012), 498.
17. Percy Shelley, 'Mont Blanc', *The Complete Poetry of Percy Bysshe Shelley*, vol. 3, 79–88, l. 1 and l. 34.
18. William Paley, *Natural Theology: Or, Evidences of the Existence and Attributes of the Deity, Collected from the Appearances of Nature* (London: J. Faulder, 1802).
19. Howard, *Climate of London*, vol. 2, vi. These words were originally published in *Athenaeum* 1 (1807): 80.
20. Ibid., 277.
21. Wood, *Tambora*, 55.
22. Howard, *Climate of London*, vol. 1, xxxiv.

23. R. W., '*On the Modifications of Clouds, and on the Principles of their Production, Suspension, and Destruction: being the Substance of an Essay read before the Askesian Society in the Session 1802*, 1803. By Luke Howard, Esq. 8vo. 2s.9d. Arch.', *Monthly Review, or Literary Journal*, 47 (1805): 432–433.
24. Luke Howard, *On the Modifications of Clouds, and on The Principles of Their Production, Suspension, and Destruction: Being the Substance of an Essay Read Before the Askesian Society in the Session 1802–3* (London: J. Taylor, 1803), 6.
25. Ibid., 4.
26. Charlotte Smith, *Charlotte Smith: Major Poetic Works*, ed. Claire Knowles and Ingrid Horrocks (Peterborough: Broadview Press, 2017), 85.
27. William Blake, *William Blake: The Complete Illuminated Books*, ed. David Bindman (London: Thames & Hudson, 2000), 102.
28. John Keats, *The Major Works*, ed. Elizabeth Cook (Oxford and New York: Oxford University Press, 2008).
29. Reno, 'Romantic Clouds', 40.
30. Howard, *Modifications of Clouds*, 3.
31. Ibid., 3.
32. Published as 'Poem on the Clouds, By Goethe', in Luke Howard, *Essay on the Modifications of* Clouds, 3rd edn. (London: John Churchill & Sons, 1865), xi.
33. Lorraine Daston, 'On Scientific Observation', *Isis*, 99.1 (2008): 98.
34. Theodore M. Porter, *Trust in Numbers: The Pursuit of Objectivity in Science and Public Life* (Princeton: Princeton University Press, 1995), 3.
35. Howard, *Modifications of Clouds*, 2.
36. Howard, *Climate of London*, vol. 1, xxxvi.
37. Howard, *Climate of London*, vol. 2, iv–v.
38. Hutton has a chapter on 'Facts in confirmation of the Theory of Elevating Land above the Surface of the Sea', in which he demonstrates the need for 'the long continued operations' of geological forces in producing land formations. His account of the immense time taken to effect significant geological change eliminates the need for creative catastrophes (and indeed ensures that what might strike a witness as a catastrophe is a minor blip in the scale of geological time). In *Theory of the Earth, with Proofs and Illustrations*, vol. 2 (Edinburgh: Messrs Cadell, Junior, and Davies, 1795), 50.
39. Howard, *Climate of London*, vol. 1, xxxv.
40. Howard, *Climate of London*, vol. 2, 291–292.
41. Ibid., 289.

CHAPTER 3

Wordsworth in the Tropics of Cumbria

Elias Greig

Striding Edge, Helvellyn
Fresh dry cloudy cold bright.
Steep frozen snow/ice ridge …
Subject [Male 44] reported to have slipped off narrow ridge.
Airlifted to hospital.
Serious—unstable fracture lumbar vertabra(e).

Long Stile
Dry Bright Sunshine.
Icy turf and snow patches …
Party of two ascending headwall above Blea Water,
subject [Male 55] stepped back to take photo slipped and fell.
Certified dead at scene …
Fatal—multiple injuries chest; multiple injuries head.

Jack's Rake, Pavey Ark, Great Langdale
Showers cloudy cool dull.
Vertical rocky stony crag …
Subject [Female] fell from popular scrambles route.
Succumbed to injuries.
Fatal—compound open fracture skull.

E. Greig (✉)
University of Sydney, Sydney, NSW, Australia

A. Collett, O. Murphy (eds.), *Romantic Climates*,
https://doi.org/10.1007/978-3-030-16241-2_3

River Eden, Carlisle
Angler, [Male 41] reported missing …
clothing and fishing tackle found beside river.
Nothing found after multi-agency search …
located more than two months later near Castletown.
Fatal—drowning.

Thornthwaite Beacon
NE-fresh/drizzle/hill fog/cool/dull.
Moderate ridge rough.
Hill Walking Party of eight
(4 adults (1 pregnant) + 4 children (aged 3, 6, 6 and 12 years))
reported themselves lost,
unable to find their way off the hill.[1]

These terse, disturbingly lyrical pieces are not, as they might appear, experimental poetry, but rather a selection from the 2012 report of the Lake District Search and Mountain Rescue Association. Compiled annually, they make grim reading. Here are picturesque tourists falling to their deaths, fishermen leaving only clothes on the shore, and families disappearing into fog; stories of fragile bodies in a hostile landscape, grotesque and fatal injuries sustained, usually, in pursuit of an apparently nourishing aesthetic experience in something called 'Nature', which, even at its most mediated, regulated by thousands of years of human settlement and centuries of aesthetic response, proves stubbornly material enough to take lives. Indeed, one of the more striking aspects of these blood-curdling incidents of death, injury, and exposure is their fundamental conformity with the incidents of death, injury, and exposure in Wordsworth's poetry. Each incident report could be the seed pearl of a Wordsworthian lyric. Lakeland has been killing, maiming, or otherwise exposing its residents and visitors in remarkably similar ways for centuries.

Against this daunting prospect of ice, mist, undertow, and precipitous drops to sharp rocks, it is interesting and oddly provoking to lay out a conventional appraisal of Wordsworthian 'Nature'—to expose a critical commonplace to the inclement weather of the landscape Romantic criticism has been so complacent about, whether for or against something called 'Nature'. In his essay of 1929, 'Wordsworth in the Tropics', Aldous Huxley provides a representative and still covertly influential summary of what he perceives as the fundamentally idealised, *domesticated* 'Nature' of Wordsworth's Lakes:

> In the neighbourhood of latitude fifty north, and for the last hundred years or thereabouts, it has been an axiom that Nature is divine and morally uplifting. ... The Wordsworthian who exports this pantheistic worship of Nature to the tropics is liable to have his religious convictions somewhat rudely disturbed. Nature, under a vertical sun, and nourished by the equatorial rains, is not at all like that chaste, mild deity who presides over ... the cozy sublimities of the Lake District.[2]

From the southern hemisphere, it is especially easy to identify the grotesque colonial attitude underwriting this statement, and Huxley's account rises to Conradian heights of irony when he claims that to travel in the tropics, in the sort of 'primitive' Nature coded, in his colonial imagination, as authentic or *real*, is 'to hack one's way laboriously through a tangled, prickly, and venomous darkness'.[3] Huxley appears to have had a terrible armchair holiday, his mental path beset with thickets of imperialist tropes.

More interesting for the purposes of this discussion is the way in which Huxley's response presages and codifies an attitude to Wordsworth's engagement with nature that remains influential: that this engagement is either naïve, bowdlerised, or conceptually and politically suspect. That is, that the 'Nature' Wordsworth engaged with was not really nature at all, but rather, in James Chandler's formulation, Edmund Burke's 'second nature' of 'use, custom, and habit'. Since 'Burke's thought' was, according to Chandler, 'the epitome of political conservatism in this period', Wordsworth's 'major work ... is conservative from the start'.[4] This 'second nature' is either consonant with, or easily transposed over the apparently idyllic landscapes of Wordsworth's poetry, where, according to Huxley, 'The god of Anglicanism had crept under the skin of things, and all the stimulatingly inhuman strangeness of Nature had become as flatly familiar as a page from a textbook of metaphysics or theology. As familiar and as safely simple.'[5]

As is so often the case, Wordsworth is here conflated with his most famous and apparently diagnostic lyric, 'Tintern Abbey', wherein he detects

> ... a sense sublime
> Of something far more deeply interfused,
> Whose dwelling is the light of setting suns,
> And the round ocean, and the living air,
> And the blue sky, and in the mind of man,
> A motion and a spirit, that impels

> All thinking things, all objects of all thought,
> And rolls through all things …
> … well pleased to recognize
> In nature and the language of the sense,
> The anchor of my purest thoughts, the nurse,
> The guide, the guardian of my heart, and soul
> Of all my moral being.

This presence felt, the poet turns to a silent Dorothy and declares that 'Nature never did betray/The heart that loved her'.[6] For Chandler, this 'greatest lyric' memorialised 'the pivotal years in Wordsworth's career'. Its famous span of 'five summers, with the length/Of five long winters', Chandler argues, charts the course of Wordsworth's initial political enthusiasm giving way to betrayal, despair, and, ultimately, reaction—towards Burke's 'second nature' and its accompanying commitment to political and religious orthodoxy.[7] Smuggled in 'under the skin of things'—the green prospect that prompts and conceals it—a cultural and political position takes on the facticity and inevitability of a physical landscape, signalling, in Huxley's words, a 'change in Wordsworth's attitude towards Nature … symptomatic of his general apostasy' from the cause of liberty.[8]

Huxley's diagnosis of Wordsworth's 'apostasy' as the result of an acute case of Pantheistic Anglicanitis is usefully 'general' and, like Chandler's narrative of transition from 'first' to 'second' nature, flattens Wordsworth's political and artistic history to Romanticism's traditional definitional parabola of aspiration and defeat, what John Bugg terms 'the excitement-to-apostasy arc'.[9] Such a flattening allows Huxley to make two implicit claims: first, that there is no mismatch or contradiction implied in yoking pantheism and Anglicanism together—the difficulty of which both motivated and haunted Samuel Taylor Coleridge for a great deal of his literary life—and, second, that Wordsworth's philosophy of 'comfortable metaphysical unreality' masquerading as Nature is consistent, clear, and directly opposed to what Huxley thinks of as a more stimulatingly alien, material nature, exoticised and orientalised into a titillating foreign darkness:

> A few weeks in Malaya or Borneo would have undeceived him. … The life of those vast masses of swarming vegetation is alien to the human spirit and hostile to it. … It is a pity he never travelled beyond the boundaries of Europe. … Europe is so well gardened that it resembles a work of art, a scientific theory, a neat metaphysical system. Its tamed and temperate Nature confirmed Wordsworth in his philosophizings.[10]

Huxley's lurid dance of archetypes and stereotypes aside, what is, perhaps, most interesting here is the manner in which the materiality of European or 'civilised' nature is neatly spelled away, become 'art', 'theory', and 'metaphysical system': that is, garden. It speaks to an ignorance of the material circumstances of rural life, necessarily yoked to elemental nature and climactic conditions, in Europe and elsewhere, which has only increased since Huxley's essay. This distance from and disregard of modes of rural production and livelihood, from what might be called working landscape, seems key to an ongoing complacency about Wordsworth's nature and, more importantly, his persistent and, for some, embarrassing materialism, what Coleridge (via William Hazlitt) described as 'a something corporeal, a *matter-of-factness*, a clinging to the palpable, or often to the petty, in his poetry'.[11]

With this complacency in mind, it is worth noting the curious similarities between Huxley's 'undeceiving' and the following passage from Alan Liu's influential *Wordsworth: The Sense of History*, a work that follows on from Jerome McGann's major intervention in the field, *The Romantic Ideology*:

> Having lived for some years in a part of Connecticut not unlike the Lake District in its reservoirs, brooks, grasslands, and deep forest dells, I would go so far as acknowledge the existence of a reservoir, brook, field, and possibly even forest (more certainly, trees). But 'nature' I have never set axe to.[12]

Huxley's imaginary wanderings through the jungle, machete in hand, and Liu's hikes through the 'deep forest dells of Connecticut' with a hatchet seem remarkably compatible: two essentially urban subjects aping tropes of struggle and survival to expose Wordsworth's apparently ideal fiction of 'Nature' as a conservative myth.

But while Huxley dismisses Wordsworth's Nature as the product of parochial naivety ('A voyage would have taught him'), Liu and others identify a more disreputable urge to suppress historical and political reference, what Marjorie Levinson, speaking, again, of 'Tintern Abbey', calls 'a concretely motivated attempt to green an actualized political prospect and to hypostatize the resultant fiction, a product of memory and desire'.[13] 'Greening', in this reading, is a key strategy in Wordsworth's avoidance of political engagements, drafted in to cover otherwise awkward moments of historical and political confrontation and contradiction with pleasant foliage. This is a mode of concealment that, in McGann's estimation, represents a fundamental characteristic of the 'poetry of Romanticism ...

everywhere marked by extreme forms of displacement and poetic conceptualization whereby the actual human issues with which the poetry is concerned are resituated in a variety of idealized localities'.[14]

Nature, reduced, in Liu's paraphrase of Andrew Marvell's 'The Garden', to 'a green thought in a green shade whose softening and distancing effect is dream stuff', is, according to McGann and others, merely the most popular of these forms, evidence of the 'increasingly dominant influence' of a Romantic ideology apparently intent on avoiding or defeating history by covering it over with prospects, fruit and flowers.[15] Thus it is that, when the poem skirts too closely to subjects and objects that might prompt questions about history and power—when the reader might take too close an interest in property rights, fences, hedges, and enclosures of all kinds—the poet directs her gaze to a daffodil, or the sky. Nature talk, according to McGann, Levinson, and Liu, like talking about the weather, is usually a running distraction or conceptual and political dodge: a mode of reification, insuring a dominant ideology against the shocking intrusions of material history.

Against this suspicious, reductivist account of 'Nature' and poetry about 'Nature', the ecocritical interventions of Jonathan Bate sought to remind us that daffodils are interesting in their own right, and that 'Nature' poetry can be an attempt at understanding natural process in a non-reductive manner: 'Postmodernity proclaims that all marks are textmarks; ecopoetics proposes that we must hold fast to the possibility that certain textmarks called poems can bring back to our memory humankind's ancient knowledge that without landmarks we are lost.'[16] Frustrated with the apparent anthropocentrism of New Historicism and its severe bifurcations of history and nature, human and non-human, Bate's ecocriticism privileges Romantic poetry in particular as a mode that posits the natural world as a place of disparate interests linked in experience, experiments in imagining other forms of being that drive towards wholeness and immersion—a prerequisite of ecological thinking. 'Scientists', writes Bate, 'made it their business to describe the intricate economy of nature; Romantics made it theirs to teach human beings how to live as part of it'.[17] In this way, weather talk can be just what it is:

> Weather is a prime means of linking spatiality and temporality—this could be why so many major Romantic poems are weather poems. A Romantic poem may be regarded as a model of a certain kind of being and dwelling … an image of ecological wholeness which may grant to the attentive and receptive reader a sense of being-at-home-in-the-world.

A dismembered, machete- and hatchet-slashed nature is here reconstituted, made whole by a criticism that, for all its self-conscious, *soil*-conscious Anglo-Saxon compounds and hyphen words, seems curiously sublimating, eager to assume a zen-like unbeing, embarrassed by its need for language, and smelling suspiciously of the incarnational fallacy:

> If writing is the archetypal place of severance—of alienation—from immediate situatedness, then how can it speak to the condition of ecological belonging? Heidegger replies with the other half of the paradox: there is a special kind of writing, called poetry, which has the peculiar power to speak 'earth'. Poetry is the song of the earth.[18]

Given the implicit mysticism of Bate's special claims for poetry, and specifically Wordsworth's poetry, it is unsurprising to find his push-back against McGann, Liu, and Levinson's hermeneutics of suspicion justified on intuitional grounds: 'I trust the intuition that locates Wordsworth firmly in nature and I reject the counter-intuitive readings which have been so influential in the academy over the past thirty years.'[19] What is surprising is the degree to which Bate's 'Ecopoetry'—which functions not as 'a description of dwelling with the earth, not a disengaged thinking about it, but an experiencing of it'—confirms such suspicions. Its phenomenological obsessions with 'motions of reverie, of solitude, and of walking are conducive', Bate claims, 'to what I shall call "ecopoetic" consciousness but', he readily admits, 'not necessarily to "ecopolitical" commitment',

> since poetic and practical language have very different purposes. … Ecopoetry is not synonymous with writing that is pragmatically green: a manifesto … will not be poetic because its language is bound to be instrumental, to address questions of doing rather than to 'present' the experience of dwelling.[20]

The Romantic Ecopoet is caught flat-footed in the perceptual moment, his commitment to 'dwelling', in line with the incarnational function of his specialist language, precluding him from 'doing' anything more directed or purposive. As Terry Eagleton notes, '[p]ushed to an extreme, this means that the truest poet would be a greengrocer. … For words which "become" what they signify cease to be words at all', disappearing 'into the objects they are supposed to denote.'[21] Confined to the aspiration of working as a clear medium or perfect mirror for experience, the poet is removed from or insulated against questions of history and politics, trapped, like an insect in amber, in 'reverie' of Nature's timeless present.

This curiously reactionary position has the effect of buttressing Huxley's, and, by extension, Chandler's, McGann's, Liu's, and Levinson's scepticism about Wordsworth's 'Nature' with an opposite but oddly complimentary idealism. Questions about the nature and purpose of Wordsworth's engagement with 'Nature' are here switched for questions about whether or not 'Nature', in and of itself, exists and whether any engagement with it can be authentic: true 'dwelling', or sublimated apostasy—a kind of green despair. This dialectical opposition plays a key role in deferring questions about the stark and often stubborn materiality of nature on show in Wordsworth's poetry, and the consequences of such starkness and stubbornness for the conception of Romanticism critics like Bate and McGann implicitly share, though from opposite sides of the question. That is, the problem of Wordsworth's materialism—his 'corporeal ... *matter-of-factness*' and 'clinging to the palpable'—is one these critics pretend to have solved for the sake of what might be termed Romanticism's critical mythology—its consistency as a period and theoretical entity—whether for or against such mythology.

Indeed, while more contemporary work by ecocritics like Timothy Morton has demonstrated how 'the idea of nature is getting in the way of properly ecological forms' of thought and enquiry, and that 'it is in art that the fantasies we have about nature take shape', these critical formulations often still rely, implicitly or explicitly, on a canonical form of Romanticism. So it is that Morton confidently identifies 'the literature of the Romantic period' as the 'target' of his 'investigation ... since it still influences the ways in which the ecological imaginary works'.[22] Morton in particular seems to operate as Bate's dark opposite, obverting Bate's verdant sublimities into a kind of hyperverbal gothicism—'How does ecognosis first appear to itself? ... The flip side of consumerism expressed in bulimia and anorexia (and punk and Wordsworth and Baudelaire) is abjection, the feeling of being surrounded and penetrated by entities that I can't peel off'—and producing a similar sense of flat-footedness: 'The level of ecological awareness after guilt and shame has to do with depression, of being depressed by the overwhelming presence of processes and entities that one can't shake off.'[23] While not at odds with Morton and contemporaries like Timothy Clark, this chapter seeks to return to the rough, sometimes waste, textual and historical grounds razed and cleared for the construction and installation of the critico-theoretical edifice of Romanticism, the pleasure dome in which the 'fantasies' of Romantic 'art' play out.[24] Excavating the Huxley/Bate binary is by no means an attempt to ignore recent developments in ecocriticism, but rather to reveal the ways in which this binary places a kind of lock on Romanticism, even in opposition.

The moments when the problem of Wordsworth's materialism presents itself most forcefully in his poetry are, like the contemporary incident reports that began this essay, moments of exposure. While Huxley and Liu read Wordsworth's nature as either a retreat from, cover for, or alternative to history, these moments of exposure, particularly in the early experimental works of the 1790s, suggest instead that Wordsworth turns to nature not as a mode of sublimation or avoidance of history, but rather as its most reliable and exacting register; what Paul Sheats calls 'a touchstone of false consciousness'.[25] Though lacking the spectacular pyrotechnic singularity of the eruption of Mount Tambora in 1815—which altered the climate, caused widespread crop failures and political upheaval, and cast a figurative and literal shadow across the lives and works of a number of writers in the so-called second generation of Romantics—the various eruptions of material nature in these works likewise provoke and provide a staging ground for historical and political engagement and critique: 'questions of doing' that fall outside the Huxley/Bate binary, exposing its limits even as they expose their protagonists to the weather.

From its outset in 1793, Wordsworth's experimental poetry traverses stark landscapes in inclement weather. Contrary to Bate's ideas about 'being' and 'dwelling', Wordsworth's first real weather poem, *Salisbury Plain*, written between 1793 and 1794, concerns itself with their dark opposites—death as opposed to being, vagrancy as opposed to dwelling. *Salisbury Plain*'s first steps carry it instantly into a landscape of endurance and dispossession, out of the walled garden of picturesque and pastoral convention into the desert of the real: 'Hard is the life when naked and unhoused'.[26] The language of this first line is, as Matthew Beaumont notes, 'Lear's language', one of many moments in the poem in which Wordsworth 'self-consciously echoes' *King Lear*.[27] Lear's peroration delivered on the heath before he takes shelter in a hovel—

> Poor naked wretches, wheresoe'er you are,
> That bide the pelting of this pitiless storm,
> How shall your houseless heads and unfed sides,
> Your looped and windowed raggedness, defend you
> From seasons such as these? (3.4.28–32)[28]

—is here struck as a sort of tuning fork, setting the pitch for what follows. Exposing himself 'to feel what wretches feel' was, for Wordsworth, made less theoretical after his own night spent 'houseless' when a carriage accident left him stranded on the edge of Salisbury Plain, with no choice but to traverse the ruin-haunted expanse on foot.

Out of this pedestrian epic a startlingly original poem would grow, the first recognisably Wordsworthian work; in Stephen Gill's words, 'a powerfully directed response to current conditions', staged on a heath-like space that is both material fact and literary artefact—an appropriation of Lear's journey among houseless heads and unfed sides and the travails of Red Cross Knight in Edmund Spenser's *The Faerie Queene*, from which Wordsworth borrows his verse form.[29] Appropriately, *Salisbury Plain* charts a traveller's journey across the Plain at sunset, the first recognisably Wordsworthian movement of a subject through a landscape:

> The troubled west was red with stormy fire,
> O'er Sarum's plain the traveller with a sigh
> Measured each painful step, the distant spire
> That fixed at every turn his backward eye
> Was lost, tho' still he turned, in the blank sky. (*SP*, 37–41)

Key to the effect of this passage is the central enjambment, 'the distant spire/That fixed at every turn his backward eye/Was lost', the fluid meeting of three lines *formally* moving the landscape even as it moves in the traveller's eyes, before being caught, 'turned', in a momentary loop between caesuras, against a 'blank sky'. Reader and traveller pass through the poem analogously, on shared poetic feet. This new sense of immediacy, of what might be called subjectival realism, 'seeks', as Sheats observes, 'to reproduce the presentation of phenomena to the mind in experience', an immersiveness that, rather than being conducive to 'reverie' and a sense of 'being-at-home-in-the-world', here produces a sense of urgency and vulnerability as the traveller casts about for shelter or direction as the light fails.[30] In a radical break from the carefully proportioned views of the picturesque, the landscape is circumscribed by the perspective of one moving through it, closing the distance between protagonist and reader, who, with no chance of retreat to a curatorial distance, is likewise exposed to the Plain's weather, presented not to be admired, but rather to be experienced and endured.

This counteractive realism is ingeniously enhanced by the deliberate inclusion of objects and tropes from picturesque and sentimental convention, blending the traveller's hopes with the reading audience's pastoral appetites. Their subsequent disappointment both amplifies the reality of the landscape and banishes and rebukes the trappings of convention:

> No shade was there, no meads of pleasant green,
> No brook to wet his lip or soothe his ear,
> Huge piles of corn-stack here and there were seen
> But thence no smoke upwreathed his sight to cheer. (*SP*, 46–49)

The repeated negatives emphasise contrast, solidifying the objects that *do* appear by the absence of those that do not, provoking what Sheats terms the 'shock of disappointment'.[31] The traveller's and audience's pastoral expectations are exhaustively dashed. *Salisbury Plain* beats back any sense of 'the pastoral felicity' John Williams identifies as occupying 'a central, reconciling role' in Wordsworth's previous works, '*An Evening Walk*, and … *Descriptive Sketches*'.[32] The single pastoral figure descried by the traveller—'And see the homeward shepherd dim appear/Far off'—soon vanishes, lost, like the convention he represents, in darkness and distance, figuratively and literally a far cry from the 'homeward plowman' of Thomas Gray's *Elegy Written in a Country Churchyard* or the 'hameward' Cotter of Robert Burns's 'Cotter's Saturday Night'.[33] He leaves the reader, the traveller, and the poem not to the luxurious crepuscular gloom of Gray's couchant narrator, or Burns's Cotter's 'clean hearth-stane' but, like Wordsworth on the brink of the Plain as Calvert rides away, to a night spent 'houseless', exposed to the elements, with a real risk of injury and every chance of becoming lost.[34]

The intentionality of this generic subversion, a tactic Sheats identifies as one of a number of 'confrontational devices in the experimental ballads of 1798 that provoke the habitual prejudices of the contemporary audience', is borne out by the appearance, in the following stanza, of a direct reference to 'Cotter's Saturday Night'.[35] As the 'homeward shepherd' disappears and the traveller looks for shelter, Wordsworth includes a verbal and visual echo of Burns's decidedly 'hameward'-pointing poem.

> Long had each slope he mounted seemed to hide
> Some cottage whither his tired feet might turn,
> But now, all hope resigned, in tears he eyed
> The crows in blackening eddies homeward borne. (*SP*, 55–60)

Burns's 'black'ning trains o' craws' pass picturesquely overhead on their way, like the Cotter, 'to their repose', emphasising, again by contrast, the traveller's and *Salisbury Plain*'s homelessness.[36] The deliberate conjuring up and shutting down of pastoral expectations, a move that, in the words of Karen Swann, 'by turns entices and reproves ... readerly appetites', deconstructs the trappings of pastoral convention by exposing what Bate terms its 'evergreen language' to the material conditions such language encodes or conceals.[37]

The political purpose of such exposure is revealed in the poem's central episode. The benighted traveller takes shelter in 'A lonely Spital' where he meets a female vagrant widowed by the American war She narrates a politically incendiary narrative of dispossession, through which Wordsworth exhaustively traces the manifold social, political, and economic forces that leave her 'houseless'. A far cry from the timeless idyll of Gray's 'sequester'd vale of life', her life is a series of disruptions and transportations, beginning with the breaking up of her childhood home.[38] Her 'father's cottage' by 'Derwent's side', complete with 'finny flood' and 'fleecy store' (*SP*, 226, 228, 233)—bland conventionalities which serve to heighten the shock when 'cruel chance and wilful wrong' (*SP*, 255) shatter its fragile self-sufficiency—is lost when her father is persecuted by a land-hungry aristocrat, revealing the precarious reality behind the idyll:

> Even to the bed where his old body lay
> His all was seized; and weeping side by side
> Turned out on the cold winds, alone we wandered wide. (*SP*, 259–261)

Desperate to get her father out of the wind, the Vagrant marries her childhood sweetheart, a weaver, 'And sheltered from the winds once more my father slept' (*SP*, 289). With the loss of her father's land, the family is entirely dependent on the young man's wage and thrown back into poverty with the outbreak of war with America, linked by Wordsworth to economic scarcity and a downturn in trade:

> For War the nations to the field defied.
> The loom stood still; unwatched, the idle gale
> Wooed in deserted shrouds the unregarded sail. (*SP*, 295–297)

As 'Labour's cheerful hum' gives way to 'Silence and Fear' (*SP*, 298–299), army recruiters 'Beat round to sweep the streets of want and pain' (*SP*,

300–301). Impelled by poverty and pride, the Vagrant's husband takes the bounty, one of many who, in the words of Joseph Gerrald, luminary of the radical London Corresponding Society,

> out of the general want ... were the more readily recruited, because the scanty pay and subsistence of a soldier became an object of envy to the unfortunate peasantry; and death seemed more honourable and less certain by the sword than by penury and disease.[39]

With her husband and children dead—one of war, the others of disease—the Vagrant is left 'homeless near a thousand homes' (*SP*, 386). Pastoral convention is here imperilled by the forces of political and economic contingency, left 'houseless' to the 'pitiless storm' poised to break out all over again as Britain went to war with France in 1793.

To expose the Vagrant's cottage to ruin in this way was to break something valuable, as a source of both conservative propaganda and literary and artistic pleasure. For writers like Hannah More and William Paley, the cottage and its associated pastoral tropes operated as sentimental spectacle and moral example. In Nigel Leask's words, 'appropriated as a conservative argument ... the cottage and the contented family life which it symbolized allayed bourgeois fears about the iniquities of poverty'.[40] Paley, in his transparently titled pamphlet of 1792, *Reasons for Contentment: Addressed to the Labouring Part of the British Public*, furnished his argument for quiescence with what Leask identifies as 'a vignette [which] except for its summery outdoors setting ... might almost be from Burns's "The Cotter"':[41]

> if the face of happiness can be anywhere seen, it is in the summer evening of a country village. Where, after the labours of the day, each man, at the door, with his children, amongst his neighbours, feels his frame and his heart at rest ... a delight and complacency in his sensations far beyond what either luxury or diversion can afford. The rich want this; and they want what they must never have.[42]

More's *Village Politics* likewise argues for contentment in a dialogue between radical Tom and honest Jack, who proclaims: 'My cottage is my castle; I sit down in it at night in peace and thankfulness. ... Instead of indulging in discontent ... I read my bible, go to church, and think of a treasure in Heaven.'[43] Key to the appeal and propagandistic value of such 'cottage scenes' was their fundamental aesthetic and moral conventionality: a political subject is replaced with an artistic object.

Keeping such depictions ahistorical and non-specific was a matter of some urgency, particularly as the effects of the war began to be felt, an anxiety carried over into the ruthless policing of temporal borders to ward off the most dangerous question asked by radical Tom: 'But the times—but the taxes, Jack'. Jack's response, despite More's best efforts, is as feeble as it is revealing: 'Things are dear, to be sure: but riot and murder is not the way to make them cheap. And taxes are high; but … I dare say, if the honest gentleman is not disturb'd by you levellers, things will mend every day.'[44] The war and its effects—'the times', 'the taxes'—made this already credulity-straining picture of things mending 'every day' thoroughly absurd. Radicals like Gerrald were quick to respond to conservative archetypes like Jack by reporting on the condition of their real-life counterparts, 'reduced to the melancholy alternative either of starving in the streets, or of taking up the trade of slaughter and rapine, and of preserving their own lives by the destruction of others'.[45]

That this passage contains the essential plot of *Salisbury Plain* is not a coincidence. Ruthlessly exposing the cherished tropes of sentimental and pastoral convention to the effects of war and oppression, Wordsworth, like Gerrald, hijacks their artistic resources, a tactic carried over into the poem singled out by Liu as 'one of the strongest cases of the denial, the overdetermined and precise absence, that is the poet's sense of history': 'The Ruined Cottage'.[46] In it, Margaret, a poor woman with two young children, is abandoned by her husband, Robert, a weaver, after the economic effects of war and a bad harvest put him out of work, forcing him to take the bounty and enlist. The extraordinary similarity of this synopsis to the events of *Salisbury Plain* hints at the fundamental correspondence and continuity between the two works—a stubbornly political mixture of war, economic downturn, and national ruin carried over into the seizure or disintegration of a cottage, situating what Sheats calls the 'psychological effects of poverty' and 'its causes … within the domestic institution … that in the 1790s offered a common synecdoche for society'.[47] Wordsworth charts the ruin of the eponymous cottage through the breakdown of its domestic economy and the incursion of livestock and wild nature which lead to its present state. In David Fairer's words, 'The Ruined Cottage' is a poem in which 'the forces of nature make demands on humans … that break the limits of the pastoral world'.[48]

Like *Salisbury Plain*, the poem begins with a narrator crossing a landscape, but with sunset swapped for high noon, and the Plain for the Lakes, tantalised with views of distant shelter, 'soft cool grass' and the 'dewy shade' of 'some huge oak whose aged branches make/A twilight of their own'.[49] This bubble of sylvan 'twilight' is suddenly burst by the combination of a caesura and end-stopped line, a dramatic prosodic break announcing a radical change of orientation to the landscape:

> ... Other lot was mine.
> Across a bare wide Common I had toiled
> With languid feet which by the slippery ground
> Were baffled still; and when I sought repose
> On the brown earth my limbs from very heat
> Could find no rest nor my weak arm disperse
> The insect host which gathered round my face
> And joined their murmurs to the tedious noise
> Of seeds of bursting gorse which crackled round. (MS. B, 17–25)

The 'cool grass' and 'dewy shade' are revealed as distant objects of longing for a narrator caught out in the heat, a complex version of the 'shock of disappointment' in *Salisbury Plain*. Contrary to the critical narrative that persistently reads the poem as a 'green thought in a green shade', the narrator, whose perspective the reader follows, moves 'towards a group of trees' (MS. B, 26) not to seek beauty but to find respite from the sun. The 'Ruined Cottage' (MS. B, 30) is not an object of aesthetic pilgrimage, but rather a remnant of habitation, transformed from an eye-involving surface of picturesque interest to the grave-like memorial of what Richard Gravil describes as a 'social' tragedy 'brought about by an unjust state of society' through the intervention of an interlocutor to whom the narrator cedes narrative authority.[50]

Armytage the pedlar is an eloquent and informed representative of the rural poor—his first line in the poem a provocative reversal of established intellectual and political hierarchies: 'I see around me ... Things which you cannot see' (MS. B, 129–130). What Armytage sees is not a vision of natural harmony; rather his 'eye' which discerns 'shades of difference ... hid in all exterior forms' (MS. B, 94–95) allows him to reconstruct what *was* from what *remains*, educating narrator and reader to do the same. From his introduction in the poem, 'stretched on a bench whose edge with short bright moss/Was green and studded o'er with fungus flowers' (MS. B, 37–38), Armytage is centred in a corona of detail, his person and narration the locus of a series of intricate descriptive images. The narrator, 'With thirsty heat oppressed' (MS. B, 110), coming within this ambit, is suddenly invested with a similar power of notice, his thirsty eye picking out Armytage's hat 'Bedewed with water-drops, as if the brim/Had newly scooped a running stream' (MS. B, 112–113). Armytage responds to the narrator's thirst by 'pointing to a sun-flower' growing over a 'wall', a natural signpost to the well that sustains it (MS. B, 114–115). Keeping the poem's clues visual rather than verbal, Wordsworth inculcates a sensitive visual acuity in the reader, a close attention prerequisite to grasping the significance of the images presented. Accordingly, as the narrator climbs the wall to slake his thirst, he discovers what K. E. Smith identifies as 'the central image-cluster of the whole poem': Margaret's garden.[51]

> ... It was a plot
> Of garden-ground, now wild, its matted weeds
> Marked with the steps of those whom as they pass['d],
> The gooseberry trees that shot in long [lank slips],
> Or currants shewing on a leafless stem
> Their scanty strings, had tempted to o'erleap
> The broken wall. Within that cheerless spot,
> Where two tall hedgerows of thick willow boughs
> Joined in a damp cold nook, I found a well
> Half choked. (MS. B, 116–125)

This garden, its well, and the cottage itself are the material registers by which Margaret's tragedy is recorded, the details and signs of which the pedlar can decipher and communicate:

> The waters of that spring if they could feel
> Might mourn ...
> ... time has been
> When every day the touch of human hand
> Disturbed their stillness, and they ministered
> To human comfort. As I stooped to drink,
> Few minutes gone, at that deserted well
> What feelings came to me! A spider's web
> Across its mouth hung to the water's edge,
> And on the wet and slimy foot-stone lay
> The useless fragments of a wooden bowl;
> It moved my very heart. (MS. B, 135–146)

While McGann and Liu both stress the poem's 'greening' as a screen of 'imagery wholly distanced from normal human concerns' to avoid or suppress political and historical questioning, nature in 'The Ruined Cottage' operates as the poem's primary mode of history, registering change and contrast.[52] Armytage's recollections of the cottage provide the poem with its plot and structure, centred, as Fairer observes, around 'the motif of the cycle or return' as Armytage revisits the cottage year on year.[53] Prompted by the wooden bowl, Armytage recalls:

> ... The time has been
> When I could never pass this road but she
> Who lived within these walls, when I appeared,
> A daughter's welcome gave me ...
> ... Many a passenger
> Has blessed poor Margaret for her gentle looks
> When she upheld the cool refreshment drawn
> From that forsaken well. (MS. B, 146–155)

Against this recollection, Armytage identifies the incursion of 'wild' nature as a key marker of the cottage's and Margaret's literal decay, evidence of the ultimate break in the poem's 'cycle', presaged brilliantly in the broken pieces of her water-bearing bowl:

> ... She is dead,
> The worm is on her cheek, and this poor hut,
> Stripped of its outward garb of household flowers,
> Of rose and jasmine, offers to the wind
> A cold bare wall whose earthy top is tricked
> With weeds and the rank spear-grass. She is dead,
> And nettles rot and adders sun themselves
> Where we have sat together while she nursed
> Her infant at her bosom. (MS. B, 157–165)

Nature functions here, in Fairer's words, 'as a kind of layered history, building up experience and memory ... a living history that is not in any way mystical or transcendent'.[54] Far from suppressing or waylaying historical enquiry, this 'layered history' of past and present, domesticity and ruin, provokes an obvious question based on the contrast between then and now: what happened?

Armytage moves to the point smartly, introducing Robert, 'an industrious man,/Sober and steady', 'busy at his loom', and plying an equally 'busy spade' in the garden, soon to be crushed by the economic effects of war, a poor harvest, and illness, backdated ten years or so to the American war, but equally descriptive of 1795–1797, the period over which the poem was composed:

> —You may remember, some ten years gone,
> Two blighting seasons when the fields were left
> With half a tillage. It pleased heaven to add
> A worse affliction in the plague of war:
> A happy land was stricken to the heart. (MS. B, 185–189)

While this imprecise dating has, Gravil notes, made 'it fashionable to criticize this poem for its marginalisation of history … frames work best when they do not obscure the pictures'.[55] In late 1794 an 'exceptionally severe winter' caused widespread 'crop failures' which, on top of the war, 'sent the price of provisions soaring', sparking what E. P. Thompson describes as an 'unprecedented rash of food rioting'.[56] Writing in 1795, William Frend noted that with the rise in the price of grain, 'every other article has risen in proportion; and the poor man's wages are not adequate to the supply of the wants of a numerous family', a situation in no way improved by 1797, when Thomas Erskine recorded 'the sinews of the laborious poor … cracking under the burdens already imposed upon all the necessities of life'.[57] Armytage's profession makes him an ideal witness to these conditions, a proxy for Wordsworth's own poetical practice over the same period:

> 'Twas a sad time of sorrow and distress:
> A wanderer among cottages,
> I with pack of winter raiment saw
> The hardships of that season: many rich
> Sunk down as in [a] dream among the poor,
> And of the poor did many cease to be,
> And their place knew them not …
> A time of trouble; shoals of artisans
> Were from their daily labour turned away
> To hang for bread on parish charity,
> They and their wives and children. (MS. B, 190–209)

Margaret's family, 'gladly reconciled/To numerous self denials', go 'struggling on through those calamitous years/With cheerful hope' until 'A fever' seizes Robert, and 'the little … stored to meet/The hour of accident or crippling age' is 'all consumed' (MS. B, 197–205). Wordsworth exposes the cottage and its associated conservative myths to the worst effects of war and economic depression. With no work and no savings, Robert is faced with terrible choices and takes the bounty, 'A purse of gold' Margaret discovers on the casement:

> Poor Man! he had not heart
> To take a farewell of me, and he feared
> That I should follow with my babes and sink
> Beneath the misery of a soldier's life. (MS. B, 323–332)

While the Female Vagrant of *Salisbury Plain* is doomed to wander, 'Robert's departure', in Gravil's words, 'fixes Margaret to the spot, in which the decay of her home … is depicted in remorselessly accumulating detail'—a series of snapshots based on the course of the pedlar's rounds, each visit showing further evidence of the incremental dissolution of the cottage and its inhabitants.[58] Leaving Margaret 'busy with her garden tools' in 'the early spring', Armytage 'measured back this road/Towards the wane of summer' to find Margaret 'absent' and cottage and garden 'changed':

> The unprofitable bindweed …
> Had dragged the rose from its sustaining wall
> And bowed it down to earth …
> And looking round I saw the corner stones,
> Till then unmarked, on either side the door
> With dull red stains discoloured and stuck o'er
> With tufts and hairs of wool, as if the sheep
> That feed upon the commons thither came
> As to a couching-place and rubbed their sides
> Even at her threshold. (MS. B, 341–342, 356–357, 361–394)

Without Robert's labour and income, the cottage is falling into decay, encroached upon by wild nature and livestock—evidence of social disregard. Unable to offer meaningful relief, the pedlar leaves, returning in spring to find Margaret, 'not changed/In person', but the change in 'her house' and garden showing 'More plainly still that poverty and grief/Were now come nearer to her' (MS. B, 438–439, 452–453). The garden, 'hard,/With weeds defaced and knots of withered grass', is haunted by images of past fruitfulness—'No ridges … of clear black mould,/No winter greenness', 'herbs and flowers … gnawed away/Or trampled'—negative presences that give way to the fundamental absence which is their source: 'Margaret stood near, her infant in her arms. … She said, "I fear it will be dead and gone/Ere Robert come again"' (MS. B, 453–465). When Armytage returns, 'beaten by autumn wind[s]', this decline is hastening—'She told me that her little babe was dead/And she was left alone'—and Margaret, still caught in a chiasmus of grief and hope, 'A wife, and widow', is increasingly desperate for news of 'him whom she had lost', a desperation that extends over 'Five tedious years' of 'unquiet widowhood' (MS. B, 474–484). Eventually the material effects of 'poverty and grief', both the result of Robert's absence, 'come' still 'nearer', and Margaret dies a lingering death of illness and exposure, the final break in the poem's cycle of returns:

> ... her poor hut
> Sunk to decay, for he was gone whose hand,
> At the first nipping of October frost,
> Closed up each chink and with fresh bands of straw
> Chequered the green-grown thatch. And so she sate
> Through the long winter, reckless and alone,
> Till this reft house by frost, and thaw, and rain
> Was sapped; and when she slept the nightly damps
> Did chill her breast, and in the stormy day
> Her tattered clothes were ruffled by the wind
> Even at the side of her own fire …
> In sickness she remained, and here she died,
> Last human tenant of these ruined walls. (MS. B, 512–528)

In charting the encroachment of wild nature on the cottage and garden, Wordsworth, through Armytage, re-stages its ruination. Far from acting as a screen or cover for history, the invasive plant life, adders, livestock, and indeed the narrator himself, who trespasses to drink, represent instead its culmination—the final result of a process that begins with the outbreak of war—from which the poem, through Armytage, reaches back.

Situating this process of historical recovery in one 'spot' revisited successively as the poem moves inexorably towards the present, Wordsworth is able to chart each change in detail, rupturing the temporal cordon erected by government apologists like More and Paley. As Fairer writes, in 'the 1790s, against a background of rapid change, disorganisation, and uncertainty, the act of revisiting is potentially a highly charged one', embarrassing conservative writers determined to pretend all was well.[59] In

1795, provoked by proposals to raise the price of labour with the price of commodities, Burke wrote to William Pitt, urging him 'manfully to resist ... breaking the laws of commerce, which are the laws of nature, and consequently the laws of God', since 'I do not know of one man, woman, or child, that has perished from famine; fewer, if any, I believe, than in the years of plenty, when such a thing may happen by accident'. As 'to the operation of the war in causing the scarcity of provisions', Burke concluded, 'I do not think it worth powder and shot', which 'ought to bind us all, rich and poor together, against those wicked writers of newspapers'.[60] In the face of this diminishment and suppression, 'The Ruined Cottage' launches a subtle but thoroughgoing retrospective on the reciprocal 'operation' of war and scarcity—backdated to the first revolutionary war of the period, a version of which Wordsworth, in company with writers like Gerrald, Frend, and Erskine, saw as being played out again in France—the result of which was national ruin, depicted in miniature in the sunken history of Margaret's cottage. The fundamental materiality of the cottage's 'change', buttressed in a physical nature unresponsive to discourse and unswayed by human reproach or appeal, takes root and breaks through—'first' nature is mobilised against 'second' nature to dispel its ideological effects.

Wordsworth's poetry of the 1790s is full of similar moments of ideological exposure, the most dramatic of which takes on the morality of nature directly. In *The Borderers*, a play completed in 1797, Herbert, a blind old man, is abducted by borderer captain Mortimer, under influence of the play's philosophical villain, Rivers, who falsely accuses Herbert of planning to sell his daughter, Matilda, to the tyrannical Lord Clifford. In mental agony, unsure of Herbert's guilt or innocence, Mortimer takes up a self-serving belief in a providential universe—'I see the end for which/ An arm invisible hath led me hither'—and abandons Herbert on a heath, in a kind of medieval trial by ordeal: 'Thou wilt have many guides if thou art innocent.' The verdict of the universe is, appropriately, sabotaged by human error: Mortimer forgets to leave Herbert his scrip containing food and water. Herbert wanders blindly over the heath towards the sound of a bell, not knowing that it rings in a ruined and deserted chapel, before succumbing to the 'stifling wind'. Matilda eventually discovers her father's corpse, from which she disturbs 'a solitary crow' given cruel significance by Herbert's retelling, immediately prior to his abandonment, of the story of Elijah, fed by ravens in the wilderness.[61] This crow, in hideous contrast, feeds on Herbert. The material nature of *The Borderers* is outside providence and directly contradicts biblical myth. Any thought of a moral universe is savagely debunked.

What to make, then, of Wordsworth's invocation of nature in 'Tintern Abbey', his claim that 'Nature never did betray/The heart that loved her', given its disruptively material character in contemporaneous works? If, as Sheats suggests, Wordsworth here 'rests truth and redemptive power in ... palpable substance that ... resists all human efforts to deny its existence', an 'anchor' or 'locus of abiding and normative values, a bedrock of truth in a void of psychic division and moral relativism', then 'the tendency of Wordsworth's development at this time is not at all commensurate with the cluster of attitudes commonly regarded as "romantic"'.[62] Materialising the nature of 'Tintern Abbey', a load-bearing text still crucial to orthodox critical and theoretical constructions of Romanticism, necessarily shifts the ground beneath this edifice, undermining its foundations, opening it to ruin. Extending the question further, how does this material nature alter established readings of other load-bearing poems by Wordsworth: the great 'Ode', with its persistently Lucretian undersong 'Of sense and outward things', or the sonnet exhorting imprisoned Haitian revolutionary, Toussaint L'Ouverture, to take comfort: 'Thou hast left behind/Powers that will work for thee; air, earth, and skies'?[63] In short, can established conceptions of Wordsworth's poetry, and the critical mythology of Romanticism itself, weather a radically material universe? Or will they likewise succumb to exposure and accident in the Tropics of Cumbria?

In any case, the determinedly material nature on show in Wordsworth's poetry of the 1790s demands an expansion of critical and theoretical parameters beyond the complimentary opposition of Huxleyan suspicion and Batean reverence. While new work by critics like Onno Oerlemans suggests 'nature as material essence was a focus of romantic attention', stressing 'a resistance to a sense that we can and should be able to come to a complete and comfortable understanding of nature, where a sense of its complexity, intricacy, and otherness is primary and sufficient', this work often seems to combine rather than transcend positions of the established dialectic, mixing Huxley's 'alien' nature with Bate's incarnational 'being and dwelling'.[64] While recognising Oerlemans's 'timely reminder of ways in which nature has the power to resist us', conceptualising nature as sublime other, or as 'a single force', risks, as Fairer points out, 'making "it" a purposeful system rather than an endlessly various process', ignoring the myriad localities in which it proves responsive 'to ingenuity and practical experience'.[65] Against this newly sublimated 'otherness', Fairer stresses that 'nature's function is not to preside over the world … but to be what it is: an infinitely various living community within whose terms and through whose energies humanity is able to survive'.[66]

For Wordsworth in the 1790s, this 'living community' was not 'a neat metaphysical system' or an ideological screen, but rather the grounds of survival, mutual aid, and solidarity itself. In this sense, the poetry of these years imaginatively coincides with the mission of the Lake District Search and Mountain Rescue Association, issuing reports of accident and exposure to commemorate loss and urge on social efforts of rescue, shelter, and relief. When record rainfall flooded Cumbria in December 2015, Richard Warren, the association's Chairman, noted that:

> During this period, over 500 team members from both within and from outside the county worked tirelessly, to help those in need of evacuation and rescue. Many worked 48 hours non-stop. … As a region we therefore remain very proud and privileged to be able to provide our voluntary rescue service. It is only possible through the generosity of those we rescue and the people visiting or living within our communities.[67]

To grasp the urgency and significance of such communal endeavours, critical conceptions of Romantic nature must be subject to further exposure.

Notes

1. *Mountain Accidents 2012* (Cumbria: Lake District Search and Mountain Rescue Association, 2012), https://www.ldsamra.org.uk/documents/LDSAMRAAnnualReport2012.pdf/, 14, 10, 46, 25, 44, 21. Line breaks added.
2. Aldous Huxley, 'Wordsworth in the Tropics', in *Do What You Will* (London: Watts & Co., 1937), 90, 92.
3. Huxley, 'Wordsworth in the Tropics', 92.
4. James Chandler, *Wordsworth's Second Nature* (Chicago and London: University of Chicago, 1984), xviii.
5. Huxley, 'Wordsworth in the Tropics', 95.
6. William Wordsworth, 'Lines written a few miles above Tintern Abbey', in *Lyrical Ballads, and Other Poems*, ed. J. Butler and K. Green (Ithaca and London: Cornell University Press, 1992), 119.
7. Wordsworth, 'Tintern Abbey', 116; Chandler, *Wordsworth's Second Nature*, xvii.
8. Huxley, 'Wordsworth in the Tropics', 101.

9. John Bugg, *Five Long Winters: The Trials of British Romanticism* (Stanford: Stanford University Press, 2014), 1.
10. Huxley, 'Wordsworth in the Tropics', 101–102, 90–91, 102–103.
11. William Hazlitt, 'My First Acquaintance with Poets', in *William Hazlitt: Selected Writings*, ed. J. Cook (Oxford: Oxford University Press, 2009), 224.
12. Alan Liu, *Wordsworth: The Sense of History* (Stanford: Stanford University Press, 1989), 38.
13. Huxley, 'Wordsworth in the Tropics', 102; Marjorie Levinson, *Wordsworth's Great Period Poems* (Cambridge: Cambridge University Press, 1986), 15.
14. Jerome McGann, *The Romantic Ideology: A Critical Study* (Chicago: University of Chicago Press, 1983), 2.
15. Liu, *Wordsworth: The Sense of History*, 314; McGann, *The Romantic Ideology*, 20.
16. Jonathan Bate, *The Song of the Earth* (London: Picador, 2000), 175.
17. Jonathan Bate, *Romantic Ecology: Wordsworth and the Environmental Tradition* (London and New York: Routledge, 1991), 40.
18. Bate, *The Song of the Earth*, 175, 251.
19. Bate, *Romantic Ecology*, 10.
20. Bate, *Song of the Earth*, 42.
21. Terry Eagleton, *How to Read a Poem* (Oxford: Blackwell, 2007), 60.
22. Timothy Morton, *Ecology Without Nature: Rethinking Environmental Aesthetics* (Cambridge, MA: Harvard University Press, 2007).
23. Timothy Morton, *Dark Ecology: For a Logic of Future Coexistence* (New York: Columbia University Press, 2016), 123, 153–154.
24. See, among others, Timothy Clark, *Ecocriticism on the Edge: The Anthropocene as a Threshold Concept* (London: Bloomsbury, 2015), and Gillen D'Arcy Wood, *Tambora: The Eruption that Changed the World* (Princeton: Princeton University Press, 2014).
25. Paul D. Sheats, 'Cultivating Margaret's Garden: Wordsworthian "Nature" and the Quest for Historical "Difference"' in *Placing and Displacing Romanticism*, ed. P. J. Kitson (Aldershot: Ashgate, 2001): 16–32, 28.
26. William Wordsworth, *Salisbury Plain* in *The Salisbury Plain Poems*, ed. S. Gill (Ithaca: Cornell University Press, 1975), 1. Subsequent references to this edition are incorporated in the text.
27. Matthew Beaumont, *Nightwalking: A Nocturnal History of London* (London: Verso, 2016), 243.
28. William Shakespeare, *The Tragedy of King Lear* in *The Norton Shakespeare*, gen. ed. S. Greenblatt (New York and London: Norton, 1997), 3.4.28–3.4.32.
29. Stephen Gill, '"Adventures on Salisbury Plain" and Wordsworth's Poetry of Protest 1795–1797', *Studies in Romanticism*, 11 (Winter 1972): 48–65, 53.

30. Paul D. Sheats, *The Making of Wordsworth's Poetry, 1785–1798* (Cambridge, Massachusetts: Harvard University Press, 1973), 88.
31. Sheats, *The Making of Wordsworth's Poetry*, 92.
32. John Williams, 'Salisbury Plain: Politics in Wordsworth's Poetry', *Literature and History*, 9.2 (1983): 164–193, 176.
33. Thomas Gray, *Elegy Written in a Country Church-Yard* in *Thomas Gray and William Collins: Poetical Works*, ed. R. Lonsdale (Oxford: Oxford University Press, 1977), 73–76; Robert Burns, 'The Cotter's Saturday Night', in *Poems, Chiefly in the Scottish Dialect* (Kilmarnock: John Wilson, 1786), 18.
34. Robert Burns, 'The Cotter's Saturday Night', 22.
35. Paul D. Sheats, '"Tis Three Feet Long, and Two Feet Wide": Wordsworth's "Thorn" and the Politics of Bathos', *The Wordsworth Circle*, 22.2 (1991): 92–100, 94.
36. Burns, 'The Cotter's Saturday Night', 13.
37. Karen Swann, 'Suffering and Sensation in *The Ruined Cottage*', *PMLA*, 106.1 (January 1991): 83–95, 93; Bate, *Romantic Ecology*, 18.
38. Gray, *Elegy Written in a Country Church-Yard*, 75.
39. Joseph Gerrald, *A Convention the Only Means of Saving Us From Ruin* (London: D. I. Eaton, 1793), 20.
40. Nigel Leask, *Robert Burns and Pastoral: Poetry and Improvement in Late Eighteenth-Century Scotland* (Oxford: Oxford University Press, 2010), 210, 213.
41. Leask, *Robert Burns and Pastoral*, 213.
42. William Paley, *Reasons for Contentment; Addressed to the Labouring Part of the British Public* (Stirling: F. Jolie, 1792), 16–17.
43. Hannah More, *Village Politics*, 3rd ed. (London: F. and C. Rivington, 1793), 18.
44. More, *Village Politics*, 10–11.
45. Gerrald, *A Convention the Only Means of Saving Us From National Ruin*, 84.
46. Liu, *Wordsworth: The Sense of History*, 325.
47. Sheats, 'Cultivating Margaret's Garden', 27.
48. David Fairer, *Organising Poetry: The Coleridge Circle, 1790–1798* (Oxford: Oxford University Press, 2009), 262.
49. William Wordsworth, 'The Ruined Cottage' MS. B, in *The Ruined Cottage and The Pedlar*, ed. J. Butler (Ithaca, New York: Cornell University Press, 1979), 9–12. Subsequent references to this edition are incorporated in the text.
50. Richard Gravil, *Wordsworth's Bardic Vocation, 1787–1842* (Houndmills, Basingstoke, Hampshire: Palgrave Macmillan, 2003), 86.
51. K. E. Smith, '"A Pile of Better Thoughts": Margaret, Silent Suffering and Silent Blessing', *Charles Lamb Bulletin*, 145 (2009): 40–50, 45.

52. Liu, *Wordsworth: The Sense of History*, 322.
53. Fairer, *Organising Poetry*, 263.
54. Fairer, *Organising Poetry*, 272, 282.
55. Gravil, *Wordsworth's Bardic Vocation*, 85.
56. E. P. Thompson, *The Making of the English Working Class* (New York: Vintage, 1966), 143.
57. William Frend, *Scarcity of Bread* (London: J. Smith, 1795), 7; Thomas Erskine, *A View of the Causes and Consequences of the Present War with France* (London: J. Debrett, 1797), 2.
58. Gravil, *Wordsworth's Bardic Vocation*, p. 84.
59. Fairer, *Organising Poetry*, p. 264.
60. Edmund Burke, *Thoughts and Details on Scarcity, Originally Presented to the Right Hon. William Pitt in the Month of November, 1795* (London: F. and C. Rivington, 1800), 32, 42–43.
61. William Wordsworth, *The Borderers*, ed. R. Osborne (Ithaca, New York: Cornell University Press, 1982), 3.3.122–3.3.123, 3.4.146, 4.1.6, 5.2.70.
62. Sheats, *The Making of Wordsworth's Poetry*, 132–134.
63. William Wordsworth, 'Ode', in *William Wordsworth*, ed. S. Gill (Oxford: Oxford University Press, 1984), 145; 'To Toussaint L'Ouverture', in *William Wordsworth*, 9–10.
64. Onno Oerlemans, *Romanticism and the Materiality of Nature* (Toronto: Toronto University Press, 2004), 5, 23.
65. David Fairer, '"Where Fuming Trees Refresh the Thirsty Air": The World of Eco-Georgic', *Studies in Eighteenth-Century Culture*, 40 (2011): 201–218, 207, 209; *Organising Poetry*, 263.
66. Fairer, *Organising Poetry*, 260.
67. *Mountain Accidents 2015* (Cumbria: Lake District Search and Mountain Rescue Association, 2015), https://www.ldsamra.org.uk/documents/LDSAMRAAnnualReport2015.pdf/, 3.

CHAPTER 4

Keats and the Poetics of Climate Change, 1816 and Beyond

Nikki Hessell

In 1816, the London botanist William Salisbury led a series of fieldtrips around London for aspiring botanists and doctors, encouraging his students to observe and collect the metropolitan flora. Salisbury reported on these outings in a series of letters in the *London Medical and Physical Journal*, 'Mr. Salisbury's Botanical Excursions'. The letters mostly consisted of a simple list of what the assembled botanists had observed, but Salisbury also felt compelled to comment on the unusual weather and its effect on the local plants. Like his fellow countrymen, Salisbury did not realize at the time that the odd weather of 1816 was a consequence of the Tambora eruption of April 1815, one of the most devastating volcanic explosions in history, which occurred in present-day Indonesia and sent a cloud of ash over the Northern Hemisphere.[1] But Salisbury was clear about the likely outcome of the season and its effects on his own area of expertise. The 'backward' spring was, he suggested, going to lead to problems with the growth and quality of plants in London. Fascinated by the research possibilities generated by this strange season, Salisbury noted at the conclusion of his 1816 series of classes that he hoped:

N. Hessell (✉)
School of English, Film, Theatre and Media Studies, Victoria University of Wellington, Wellington, New Zealand
e-mail: nikki.hessell@vuw.ac.nz

A. Collett, O. Murphy (eds.), *Romantic Climates*,
https://doi.org/10.1007/978-3-030-16241-2_4

> to resume this pleasant task next April, at which time I anticipate the pleasure of comparing the different periods of blooming of plants,—this being, in my opinion, the most backward season that has ever been noticed, so much so, that there will scarcely be time for the acorns or chestnuts to grow to their usual and full size. Several of the gentlemen who have honoured me with their attendance this summer, and who are now settled in different parts of the country, have obligingly offered to compare notes, on this head, with my class next season.[2]

Many of the students who had witnessed these phenomena alongside Salisbury, and who had been encouraged to observe the effects of the climate on plants in such minute detail, were trainee doctors at Guy's Hospital. Salisbury's advertisement for his course emphasized the contemporary significance of botany as a medical discipline, '[i]t having become indispensably necessary for students in medicine to obtain a knowledge of this science, or at least so much of it as relates to an acquaintance with all the plants described in the materia medica'.[3] The announcement of this new course in botany would have been of particular interest to one student at Guy's: John Keats.

Keats had begun his studies at Guy's in 1815, following a medical apprenticeship with Thomas Hammond, and would spend most of the next two years training to be a surgeon apothecary. His programme of study, as Hermione de Almeida, R. S. White, and Donald C. Goellnicht have demonstrated, included courses in chemistry, medical theory, anatomy and surgery, general medical practice, and Salisbury's botany course. The latter course, which was based on a series of practical 'Herborising Excursions' in neighbourhoods on the fringes of London, was divided into two offerings in 1816: a spring version that ran from 23 April to 23 July and a summer version from 3 August to 24 September.[4] Keats would have taken the spring course, finishing it in time to be examined for the Licentiate of the Society of Apothecaries certificate on 25 July, an examination that required detailed botanical knowledge in order to answer questions about the *materia medica*.

Keats's poetry from this year serves as testimony to his botanical studies. The verses of autumn and winter 1816, many of which were included in the 1817 *Poems*, have been analysed in detail for the precision of their botanical knowledge by de Almeida, Goellnicht, and Gareth Evans, while Nicholas Roe, Jeffrey Cox, Porscha Fermanis, and Alan Bewell have provided authoritative readings of the political and literary contexts of the natural imagery in the volume.[5] What has escaped attention, however, is the connection between these poems and the wider question of the peculiar climate of 1816. Keats was outside, observing plants, in the strange spring of that year, and the botanical elements of the 1817 *Poems* can be read as part of a long-term engagement with climate that continued well beyond 1816, but which was decisively shaped by his experiences with Salisbury.

My argument here relates to two recent, related, and inevitable developments in literary critical discourse: Romantic scholarship's turn to climate change studies and climate change activism's turn to Romanticism. The first of these turns informs the majority of my chapter. It is twenty years since Jonathan Bate, one of the foundational thinkers of ecocriticism, asked of the ode 'To Autumn': '*why* does Keats imagine autumn thus in the autumn of 1819?' and answered his own question with data about the Tambora eruption, the terrible weather of 1816 and its disastrous effects on food production, and the subsequent stabilizing of the climate in 1819.[6] The role of climate change in Keats's work has been taken up more recently by Eric Gidal, who links Keats's poetry decisively to a wider Romantic climatology and indeed to a twenty-first-century perspective on humans and the environment.[7] My aim in this chapter is to occupy the space between Bate's historicized and Gidal's globalized account. Keats's aesthetic engagement with climate in the major odes is certainly shaped by the return of a stable climate in 1819, but it also draws definitively on his experiences—poetic and botanical—in 1816, the 'Year Without a Summer'. Bate's argument about 1819 and the major odes can thus be augmented by detailed attention to what Keats learnt about his craft and his climate in the period in which he drafted some of his earliest published poems. Towards the end of this chapter, I will also consider the ways in which Keats's reflections on climate change can be felt in contemporary activist poetry.

Keats's engagement with climate is evident from the very first lines of the opening poem in the 1817 volume, 'I stood tip-toe upon a little hill', which he completed in late 1816:

> I stood tip-toe upon a little hill,
> The air was cooling, and so very still,
> That the sweet buds which with a modest pride
> Pull droopingly, in slanting curve aside,
> Their scantly leaved, and finely tapering stems,
> Had not yet lost those starry diadems
> Caught from the early sobbing of the morn. (1–7)[8]

This opening description belongs to the backward season of 1816. The 'scantly leaved' plants of that year were noted by Salisbury himself, who commented, with evident surprise, that the oaks at Combe Wood, which the class had seen on Friday 24 May, were 'as perfectly naked as they were at Christmas'.[9] The atmospheric conditions depicted in these lines are perhaps just poetic commonplaces that conjure up the sense of a morning, but the description of the air as 'very still' also suggests something unusual about the weather, something that requires close observation and comment. The 'cooling' air of Keats's lines likewise connects them with the effects of the Tambora eruption, which caused 1816 to be the Northern Hemisphere's second-coldest year since 1400.[10] In this context, we might wonder about the direction in which the verb 'cooling' sends us: is the air 'cooling' the speaker, transitively, or is it just 'cooling', intransitively, gradually becoming colder? Keats's speaker appears to be outdoors in the morning, when the dew—'the early sobbing of the morn'—is still on the leaves of the plants he observes. Salisbury's class met at 10 a.m., and thus Keats's speaker is like a member of that troupe, heading out to botanize.[11] The day might be expected to be slowly warming up, but the climate of the poem is 'backward', as Salisbury described it.

The effects of the backward climate are evident whenever Keats's speaker engages in the necessary close observation in this poem, such as the often-discussed evocation of a 'tasteful nook':

> A bush of May flowers with the bees about them;
> Ah, sure no tasteful nook would be without them;
> And let a lush laburnum oversweep them,
> And let long grass grow round the roots to keep them
> Moist, cool and green; and shade the violets,
> That they may bind the moss in leafy nets.
>
> A filbert hedge with wildbriar overtwined,
> And clumps of woodbine taking the soft wind
> Upon their summer thrones; there too should be
> The frequent chequer of a youngling tree,
> That with a score of light green brethren shoots
> From the quaint mossiness of aged roots[.] (29–40)

This passage sounds like the kind of over-full natural scene that came to typify Keats's verse, but it is in fact oddly empty and provisional. 'And let' laburnum and grass grow round the May flowers, Keats's speaker implores, just as he requests shade for the violets 'that they *may* bind the moss'. These are not conditions that the speaker currently observes, but those that he hopes will occur in the future in order for the plants to flourish. Even the initial prompt for these reflections, the 'bush of May flowers', which seems to be a stable, concrete phenomenon, is undermined by the sense in the following line that it might be possible for a garden nook to be 'without them'.

The bees form another link to the climate in 1816: *The Times* had reported in October of that year that the bees had been 'prevented by the wetness of the late summer from pursuing their labours abroad, in consequence of which many of them are too poor to subsist through the winter'.[12] William Wordsworth wrote to an acquaintance in 1817 that there was almost no honey available in England, 'last summer having been so extremely wet that the Bees were incapable of working, and most of them perished through hunger in the course of the Winter'.[13] These strangely absent bees buzz their way through Keats's 1817 *Poems*, featuring as part of a rhetorical question in 'Sleep and Poetry' (2–4) and as emblematic of summer in poems that deal with memories (such as 'To My Brother George' [13–14]), imaginary scenes (as in 'Calidore: A Fragment' [66–68]), or future experiences ('O Solitude! if I must with thee dwell' [8]), but not present realities. In 1816, the bees were not there to perform the work, botanical or poetic, that they might otherwise have done. They are the ultimate symbol of the worrying climate change of that spring.

The ecosystem Keats so carefully describes in his description of the 'tasteful nook' faces at best an uncertain future and might in fact not exist at all in the poem's moment. This possibility could add weight to readings of Keats's nature as largely ornamental and artificial, or, as Andrea Henderson has argued in relation to the 'nook' passage, as resolutely domesticated and suburban, but there is some solid scientific evidence for the necessity of imaginary flowers in 1816.[14] Salisbury offered the *London Medical and Physical Journal* a table of the plants observed on his excursion of 23 April 1816, one in which Keats may well have participated, along with the comment that 'the following scanty list of plants was all that we met with, and consequently all that was in bloom at that season'. On the list, Salisbury notes, again with apparent surprise, that the cow parsley was 'but just out', the white birch 'scarcely out', and the ladies' smock visible with 'scarcely a petal Open' (May 1816: 430).[15] These are Keats's May flowers, and they are indeed fragile and provisional because of the cold spring.

'I stood tip-toe' is a poem that foregrounds its own observational techniques, describing the speaker's intention to act as both poet and botanist and 'watch intently Nature's gentle doings' (63). The observational nature of the poem has long been acknowledged and forms the cornerstone of Marjorie Levinson's influential reading, which rightly declares that '[t]he primary action of ['I stood tip-toe'] is, of course, looking'.[16] Keats's observational technique, she argues, leads to 'the peculiarly abstract effect of these terrifically detailed inventories'.[17] This sense of abstraction comes from the provisionality or artificiality of those inventories, the paradox that the lushly evoked flowers never really convince us. Levinson interprets this as a function of class, but might it in fact manifest some of the precision of Keats's observational skill? There is another way to triangulate 'looking', 'detailed inventories', and 'abstract effect'. As Leigh Hunt astutely noted in his review of the volume, Keats's 1817 *Poems* contained 'a superabundance of detail' derived from a 'microscopical' method.[18] This combination of method and its consequent poetic detail can be found by exploring Keats's experiences in 1816, in which he was required to observe nature closely and note how much was missing, before generating a lush poetic world to fill the vacancy when he came to rewrite those experiences as poetry. Hunt was unintentionally apt when he described the detail of these poems as 'unseasonable', a term which draws attention to the strange season in which poems like 'I stood tip-toe' were composed.

An awareness of Keats's training and the seasonal peculiarity of 1816 helps to explain the way in which vision itself is depicted in the poem, particularly in lines 15–22:

> There was wide wand'ring for the greediest eye,
> To peer about upon variety;
> Far round the horizon's crystal air to skim,
> And trace the dwindled edgings of its brim;
> To picture out the quaint, and curious bending
> Of a fresh woodland alley, never ending;
> Or by the bowery clefts, and leafy shelves,
> Guess where the jaunty streams refresh themselves. (15–22)

To Levinson, the tentative verbs—'To peer about … And trace … To picture out … Guess where'—speak of 'the prurience of the project'.[19] Keats is engaged, in Levinson's reading, in an attempt to grasp at both an experience and a poetic perspective that are inaccessible to him. But the tentativeness of the verbs also links to the wider challenges facing the botanist in 1816: the struggle to discern what should have been obvious in an English spring, and the fascination, manifested in Salisbury's accounts of his classes, with recording that struggle.

The 1817 volume is filled with such unstable, intangible natural scenes. From the rhetorical questions about a lush summer tableau that open 'Sleep and Poetry' (1–8), to the 'warm desires' (85) of the speaker 'To see' (86, 92), 'To mark' (90), and 'To feel' (91) the effects of sunshine in 'To Charles Cowden Clarke', the 1817 *Poems* is packed with detailed images of a summer that is in fact entirely hypothetical. Provisional language abounds in these descriptions, working at productive odds with the layers of rich imagery, to produce a mixture of the conditional and the cornucopian. Keats's nature in 1816 is scientifically analysed, richly detailed, ostensibly observable, but only figuratively present. It is the nature of a botanist, working in what he calls in 'O Solitude! if I must with thee dwell' 'Nature's observatory' (4), who finds himself confronted with a botany of absence.

Keats's thinking about climate in 1816 was clearly connected to his thinking about poetry. Nature, in these poems, is not simply a trope, nor an aspect of the imagery, nor a conventional gesture towards the English poetic tradition. In the unstable, ephemeral form in which Keats encountered it in 1816, nature is instead a living analogy for the fragility of poetry. But, as well as its metaphorical usefulness, it offers Keats a kind of productive uncertainty, in which the absence of literal plants and insects creates room for—and perhaps necessitates—the engagement of the poetic faculties to generate, plant, and fertilize a flourishing space. After articulating his 'warm desires' for a ripe natural scene in the lines from the epistle 'To Charles Cowden Clarke' quoted earlier, Keats writes:

> No sooner had I stepp'd into these pleasures
> Than I began to think of rhymes and measures: (97–98)

These lines operate in two ways. They might lead us to question the callow speaker, who can only think about poetry when he has literally stepped into a scene which, the poem tells us, he does not and may never be able to inhabit. If the requirements of composition demand attaining such a scene, then the speaker's poetic career is as imaginary as the scene itself. But the lines also suggest that there is no literal step to be taken here. What the speaker implies is that the process of imagining such a scene provokes an outpouring of poetry. The absence of the scene is itself generative of the verse, just as the absent flowers and bees of 1816 demand a poetic response. Something new can grow out of—can only grow because of—the cold soil of the spring.

These are lessons that Keats took with him throughout his life and which were manifested with much greater poetic skill in his mature poetry, which is redolent with imagined gardens and reflections on the ways in which verse supplants specific natural phenomena. His *annus mirabilis* coincided with the return to stable temperatures and the normal progression of the seasons in England.[20] Keats noticed the change, writing to his sister Fanny that '[t]he delightful Weather we have had for two Months is the highest gratification I could receive—no chill'd red noses—no shivering—but fair Atmosphere to think in'.[21] This letter was written on 28 August 1819, and it is noticeable that, despite it being the height of summer, Keats is surprised to find that he has not been 'chill'd' or 'shivering'. Like this letter, the major odes are deeply concerned with implied comparisons between the seasons of 1816 and 1819, and indeed between the poet of 1816 and that of 1819, as Keats maintains the mental link he had established between close natural observation and the production of poetry. Ideas that had been rendered scientifically in his early botanical thinking are replayed more lyrically in the later poems, but the lessons of the botanizing process are never entirely lost.

One throwaway comment in a letter to J. H. Reynolds can help to illuminate Keats's development between 1816 and 1819 with regard to botany and poetry. Writing from Oxford in the autumn of 1817, Keats declared:

> I have not time to elucidate the forms and shapes of the grass and trees; for, rot it! I forgot to bring my mathematical case with me; which unfortunately contained my triangular Prism so that the hues of the grass cannot be dissected for you—.[22]

Keats jokes here about a stereotypical scientific mentality, in which the lack of the correct equipment makes it impossible for him to give any detail about something that he might simply have observed with his own eyes and described with his own imagination. The humour is self-aware, signalling a move away from this mentality and awareness of some of its synesthetic absurdity, but still acknowledging a pre-existing personal connection to the discipline of botany. He has literally (and perhaps metaphorically) left behind his botanical toolkit, but he still owns one, and he suggests, however self-deprecatingly, that an initial scientific observation must happen before rhapsodic poetry on the same topic can occur.

'Ode to Psyche' was composed in April 1819, before it became clear that the summer and autumn were to be pleasant and settled. The poem epitomizes the ways in which the lessons of 1816—climatic, botanic, and poetic—have come to shape Keats. The mental landscape that ends 'Ode to Psyche' is, in essence, a more sophisticated version of the botanical poems of 1816 and 1817. Several traces of 'I stood tip-toe' can be felt in this mature poem. The description of the wide open scene—'Far, far around shall those dark-cluster'd trees / Fledge the wild-ridged mountains steep by steep' (55)—is followed by the closely observed, domesticated 'rosy sanctuary' (59), in an echo of the move from the view of the horizon to the 'tasteful nook' in 'I stood tip-toe'. The 'buds, and bells, and stars' (61) are all imagined, as are the bees (56). The lack of specificity in these depictions suggests their dreamlike quality, but also, in the context of the events of 1816, begins to sound vaguely ominous. The 'breeding flowers', which Fancy 'will never breed the same' (63), are intended to signal a fruitful, varied imagination, which generates new and creative possibilities at every turn. But the same line also carries hints of thwarted organicism and threatened extinction; the flowers will, perhaps, 'never breed the same' as they once did. The stars 'without a name' (61) can be thought of as the speaker's own invention, but can also stand for a lost botanical knowledge, made more apocalyptic by the connection between star-shaped flowers and stars as cosmic bodies. The ecosystem that ends 'Ode to Psyche' has taken a drastic, potentially decisive turn towards oblivion.

But out of this looming oblivion comes a major development in Keats's poetry. While it is evident to any reader that this landscape is a metaphor for the speaker's own thought processes, rather than simply a poetic depiction of an imagined garden, reading 'Ode to Psyche' alongside the earlier botanical poems adds nuance to this interpretation. The speaker's impulse is not simply to replace those pines with 'branched thoughts' (52) and thus make use of a convenient metaphor. It is rather to acknowledge the absolute absence of plants and the necessity of imagining them, and then to reflect on the ways in which poetic thought can take root and flower in those empty spaces. The mature Keats reaches beyond the insights of his early poems, which manifest the speaker's ability to imagine plants where none existed, and instead writes the effects of this thought process into the poetry itself. The questions raised by earlier poems, about the relationship between poetry and nature in an unstable natural world, are answered here with a certainty tinged with melancholy: if one cannot count on the regeneration of nature, one's poetic work will have to become generative. In 'Ode to Psyche', Keats's own thoughts are nurtured in a soil that is available precisely because it is not sustaining plants, and the poet can thus occupy space that was previously the realm of the botanist.

By May 1819, as Keats composed 'Ode to a Nightingale', the weather had improved further and he was in a position to conduct the sort of seasonal comparison that Salisbury had urged his students to undertake. In an echo of the remark to Reynolds that 'I have not time to elucidate the forms and shapes of the grass and trees', the speaker of 'Nightingale' notes 'I cannot see what flowers are at my feet' (41). The skill of botanical observation is deployed here via the sense of smell rather than sight, as the speaker is forced to 'guess each sweet' from their scents, but the same detailed inventory of flowers is nevertheless obtained and recorded:

> White hawthorn, and the pastoral eglantine;
> Fast fading violets cover'd up in leaves;
> And mid-May's eldest child,
> The coming musk-rose. (46–49)

But these scents are those 'Wherewith the seasonable month endows / The grass, the thicket, and the fruit-tree wild' (44–45). What, we might ask, would occur in an unseasonable month? Keats was composing this poem in May 1819; three years earlier, in the same month, he would have been botanizing with Salisbury. Some of the plants Keats mentions here were behaving very strangely in that earlier moment. Salisbury, reporting on his 24 May excursion with his students, commented that 'The Hawthorn is not in bloom that I have yet seen'.[23] On 11 June, he noticed that 'The Hawthorn in the hedge in Battersea Fields, where we collected the specimen on the 31st of May, is in great beauty, but the bloom is not fully expanded'.[24] May 1819 might well have been a 'seasonable month', as temperatures and the seasons stabilized, but the necessity to draw attention to the apparently natural correlation between seasons and months serves as a reminder of the instability of the climate at the opening of Keats's poetic career.

The return of a stable climate was clear by September 1819, when Keats wrote 'To Autumn'. Here at last was a 'seasonable' season, behaving as it ought. This realignment in part accounts for the immersive, sensuous tone of the poem, as Keats revels in an orthodox autumn, so unlike the autumn of 1816; as he wrote to J. H. Reynolds, 'How beautiful the season is now—How fine the air. A temperate sharpness about it. … I never lik'd stubble fields so much as now—Aye better than the chilly green of the spring. Somehow a stubble plain looks warm—in the same way that some pictures look warm—this struck me so much in my sunday's [sic] walk that I composed upon it'.[25] As Bate has argued, the context of the climate must shape the way we read the poem Keats began composing on this walk, right down to some very specific facts about the weather:

> [O]ur understanding of 'To Autumn' should begin with the knowledge that the weather was clear and sunny on 38 out of the 47 days from 7 August to 22 September 1819, and that in the week of 15–22 September temperatures were in the mid-sixties, whereas in the corresponding week in each of the three previous years they had been in the mid-fifties. Remember the meteorological and consequent agricultural pattern: the terrible summer and failed harvest of 1816, bad weather and poor harvests continuing in 1817 and 1818, then at last in 1819 a good summer, a full harvest, a beautiful autumn.[26]

The first stanza of 'To Autumn' stresses that the natural processes are underway in a season characterized by 'fruitfulness' (1). This fruitfulness borders on excess: autumn 1819 has 'set budding more, / And still more, later flowers for the bees' (8–9), the repetition suggesting a climate that is in overdrive, making up for lost time. These bees are, of course, an especially telling sign of the recent climatic volatility. Their behaviour in 'To Autumn' is characterized by hyperactivity, with Keats depicting them as creatures who 'think warm days will never cease' (10), and whose natural activity is joyously out of kilter, 'For Summer has o'er-brimm'd their clammy cells' (11). These bees are still feeling the effects of 1816, and it seems that, in his 1819 poems, Keats is too. In this context, his famous question 'Where are the songs of Spring? Ay, where are they?' (23) reminds us of the 'backward' season he spent with Salisbury, in which spring was sought in every flower and tree, but never fully located.

The lessons of 1816 contain so much of what we now consider to be the Keats aesthetic. Deferred enjoyment, provisionality, melancholia, and sensuousness have been the keynotes of reading Keats from the earliest, most virulent criticism he received during his lifetime to the more sympathetic interpretations of modern scholarship. All these attributes were drawn out of Keats in the spring of 1816, not simply because he was studying botany but because he was studying it at a time when the climate radically disrupted the norms of natural growth and development. Keats's own growth and development is intimately connected with that season, in which it finds its mirror and its motivation. But it might also be argued that there was no way for Keats to blossom poetically until the climate stabilized. Richard Macksey has made a similar point about the ways in which 'To Autumn' reflects a 'temperate climate of the imagination', but there is a literal climate change to consider here too.[27] Keats's best poetry was not produced in the grip of the 'Year Without a Summer', nor in the following year when it was possible that cold springs and barren plants were the new normal. It took the 'delightful weather' of 1819, with its seasonal certainty, to allow a mature reflection on the productive uncertainty of previous years. Only with that stable, seasonal calendar in place could the poetic flowers of 1816 start to bloom.

That same extraordinary season, and the poetry it provoked, might also help us to understand undercurrents in contemporary poetry about global warming. In 2015, *The Guardian* published a series of twenty new poems about climate change, entitled 'Keep it in the Ground: A Poem a Day', curated by the UK Poet Laureate Carol Ann Duffy.[28] Explicit references to Romanticism are everywhere in these verses; Rachael Boast's poem 'Silent Sea' takes its title and epigraph from 'The Rime of the Ancient Mariner', while Sean Borodale's 'Scratching for Metaphor in the Somerset Coalfields' also considers Coleridge, referring to the poet's time in Somerset in an imaginary encounter with the coal mines:

> Radstock. 1794. The Fever of August.
> Coleridge is crossing a boundary to his lyric field;
> by counter-spirit. Under his feet
> Old Pit is open: boys and men
> mine its difficult, faulted, folded vein in the dark;
> their candles opening limited allowance of light. (6–11)

These examples take the characteristics of Coleridgean Romanticism as useful touchstones for modern climate discourse. Keats's influence on this series is, I would argue, more subtle, perhaps even unconscious. It can be felt in Don Paterson's 'beeless glade' (9) in his poem 'Nostalgia', which deals with a flooded landscape, or in Michael Longley's description of sitting in 'An imagined underground' formed by 'a leafy canopy' (11–12) in 'Storm', his account of a storm-damaged suburban house and garden. Most obviously, alongside other influences such as that of Gerard Manley Hopkins, it can be felt in parts of Gillian Clarke's 'Cantre'r Gwaelod':

Grieve for lost wilderness—for the lovesick salmon,
lured by sweet river-water sleeved in the salt,
homing upstream to spawn at the source
where it was born; for mating hares
in love with the March wind; for thermals
lifting a flaunt of red kites over the wood;

for bees mooning for honey in weedless fields;
for sleepy Marsh Fritillary butterflies
swarming the ancient bog of Cors Llawr Cwrt;
for the Brown Hairstreak in love with blackthorn
and the honeydew of aphids in the ash;
for the blackbird's evening aria of possession;
for Earth's intricate engineering, unpicked
like the flesh, sinews, bones of the mother duck
crushed on the motorway, her young
bewildered in a blizzard of feathers;
the balance of things undone by money,
the indifferent hunger of the sea. (19–36)

Like Keats, Clarke produces an overloaded inventory of imagery: not a butterfly, but a 'Brown Hairstreak'; not birds, but 'red kites'; not a plant, but 'blackthorn'. Like Keats, Clarke tracks the changing year through a host of natural phenomena, whose behaviour marks out the seasons: the spawning salmon of autumn and the mating hares of spring, the 'sleepy' Marsh Fritillary butterflies preparing for winter hibernation and the amorous Brown Hairstreak butterflies laying their summer eggs on the blackthorn. And like the Keats of 1816, Clarke produces a portrait of a nature that is not there to be observed in the detail with which she observes it; the poem's footnote reads 'The storms of 2014 uncovered a petrified forest and evidence of ancient habitation from the beach at Borth', to which Clarke has responded by imagining the 'lost wilderness' that the forest housed.

Keats's life was short, but he did live long enough to see the best version of 'climate change', the stabilization of the global climate following a catastrophic, though natural, disruption. 'To Autumn' celebrates this return, a present moment described in rich, sensual language. Clarke's vision of the former life of the petrified forest sounds very like 'To Autumn' in its eroticism, its lush imagery and diction, and even in some small details, such as the wind lifting the red kites, with its echo of Keats's 'winnowing wind' (15) that lifts Autumn's hair, or the sounds of the birds at evening. But for Clarke and her epoch, this is a vision of a world entirely lost, a world for which we must now '[g]rieve' (19). If there is to be a restoration, it is not yet here to be observed and celebrated. Keats's poem ends with birdsong:

> ... with treble soft
> The red-breast whistles from a garden-croft;
> And gathering swallows twitter in the skies. (31–33)

Clarke also has a birdsong, 'the blackbird's evening aria of possession' (30), but the poem cannot conclude there. Instead, we end with an image of another bird crushed, mutilated by modernity, silenced.

What does this mean for the poetry of climate change today? The writers included in *The Guardian*'s collection might not live to see a stable climate; in fact, a stable climate might never return. But if contemporary poets will not get a moment of relieved and settled reflection like that encapsulated in 'To Autumn', they can still profit from the same productive uncertainty that shaped the 'Ode to Psyche' and prompted Keats's call for a generative poetics in the face of a barren world. The poems that Duffy curated already bear the hallmarks of that productive uncertainty. They are elegiac in tone, yet mourn something that has not yet passed. Here, then, is a new and urgent genre for our times, borne out of that same generative poetics: an elegy for the living, a warning elegy, composed to remind us of what is not yet dead and to urge us to act before its words become necessary. The poetry of climate change has its music too, and it is a poetry that might prove to be our salvation.

Notes

1. The literary and cultural context of the eruption is covered by Gillen D'Arcy Wood *Tambora: The Eruption that Changed the World* (Princeton: Princeton University Press, 2014), and William K. Klingaman and Nicholas P. Klingaman, *The Year Without a Summer: 1816 and the Volcano that Darkened the World and Changed History* (New York: St. Martin's Press, 2013). A denser scientific discussion of the eruption and its aftermath can be found in the essays and case studies in C. R. Harington ed., *The Year Without a Summer? World Climate in 1816* (Ottawa: Canadian Museum of Nature, 1992).
2. William Salisbury, [Letter], *London Medical and Physical Journal* 36.4 (October 1816): 348–350, 349.

3. 'Medical and Philosophical Intelligence', *London Medical and Physical Journal* 34.3 (September 1815): 258.
4. The dates of Salisbury's classes can be reconstructed from the letters he wrote to the *London Medical and Physical Journal*.
5. See Hermione de Almeida, *Romantic Medicine and John Keats* (New York and Oxford: Oxford University Press, 1991), 146–174; R. S. White, 'Keats and the Crisis of Medicine in 1815', *Keats-Shelley Review* 13 (1999): 58–75; Donald C. Goellnicht, *The Poet-Physician: Keats and Medical Science* (Pittsburgh: University of Pittsburgh Press, 1984), 85–97; Gareth Evans, 'Poison Wine—John Keats and the Botanic Pharmacy', *Keats-Shelley Review* 16 (2002): 31–55; Nicholas Roe, *John Keats and the Culture of Dissent* (Oxford: Clarendon Press, 1997), 116–128; Nicholas Roe, 'John Keats's "Green World": Politics, Nature and the Poems', in *The Challenge of Keats: Bicentenary Essays 1795–1995*, ed. Allan C. Christensen, Lilla Maria Crisafulli Jones, Giuseppe Galigani and Anthony L. Johnson (Amsterdam: Rodopi, 2000), 61–77; Jeffrey N. Cox, *Poetry and Politics in the Cockney School: Keats, Shelley, Hunt and their Circle* (Cambridge: Cambridge University Press, 1998), 78–122; Porscha Fermanis, *John Keats and the Ideas of the Enlightenment* (Edinburgh: Edinburgh University Press, 2009), 27–38; Alan Bewell 'Keats's Realm of Flora', *Studies in Romanticism* 31.1 (1992): 71–98.
6. Jonathan Bate, 'Living with the Weather', *Studies in Romanticism* 35.3 (1996): 431–37, 436.
7. Eric Gidal, '"O Happy Earth! Reality of Heaven!": Melancholy and Utopia in Romantic Climatology', *Journal for Early Modern Cultural Studies* 8.2 (2008): 74–101, 95–97.
8. All references to Keats's poetry are to John Keats, *Complete Poems*, ed. Jack Stillinger (Cambridge, MA: Belknap Press, 1982).
9. William Salisbury, [Letter], *London Medical and Physical Journal* 35: 6 (June 1816): 516–517, 517.
10. Klingaman and Klingaman, *The Year Without a Summer* 20. For more detailed scientific analysis of the data about the European climate in 1816, see John Kington, 'Weather Patterns over Europe in 1816', in *The Year Without a Summer? World Climate in 1816*, ed. C. R. Harington (Ottawa: Canadian Museum of Nature, 1992), 358–371; and K. R. Briffa and P. D. Jones, 'The Climate of Europe during the 1810s with Special Reference to 1816', in *The Year Without a Summer? World Climate in 1816*, ed. C. R. Harington (Ottawa: Canadian Museum of Nature, 1992), 372–391.
11. William Salisbury, [Letter], *London Medical and Physical Journal* 35.5 (May 1816): 430.
12. [Untitled news report], *The Times* (12 October 1816): 3.

13. William Wordsworth, *The Letters of William and Dorothy Wordsworth: The Middle Years, Part II*, 2nd ed. Ernest De Selincourt, rev. Mary Moorman and Alan G. Hill (Oxford: Clarendon Press, 1970): 386.
14. Andrea Henderson, 'Mastery and Melancholy in Suburbia', *18th Century: Theory and Interpretation* 50: 2–3 (2009): 221–244, 236. See also Bewell, 'Keats's Realm' 71–98; and Elizabeth Jones, 'Keats in the Suburbs', *Keats-Shelley Journal* 45 (1996): 23–43.
15. William Salisbury, [Letter], *London Medical and Physical Journal* 35.5 (May 1816): 430.
16. Marjorie Levinson, *Keats's Life of Allegory: The Origins of a Style* (Oxford: Basil Blackwell, 1988), 240.
17. Ibid., 240.
18. Quoted in G. M. Matthews, ed., *Keats: The Critical Heritage* (London: Routledge and Kegan Paul, 1971), 58.
19. Levinson, *Keats's Life* 239–240, ellipsis in the original.
20. Wood, *Tambora* 39.
21. John Keats, *The Letters of John Keats, 1814–1821*, 2 vols, ed. Hyder Edward Rollins (Cambridge: Cambridge University Press, 1958), 2: 148.
22. Ibid., 1: 162.
23. William Salisbury, [Letter], *London Medical and Physical Journal* 35.6 (June 1816): 516–517, 517.
24. William Salisbury, [Letter], *London Medical and Physical Journal* 36.1 (July 1816): 84–86, 84.
25. Keats, *Letters* 2: 167.
26. Bate, 'Living' 440.
27. Richard Macksey, '"To Autumn" and the Music of Mortality: "Pure Rhetoric of a Language without Words"', in *Romanticism and Language*, ed. Arden Reed (Ithaca: Cornell University Press, 1984), 263–308, 265.
28. The complete collection can be found at https://www.theguardian.com/environment/series/keep-it-in-the-ground-a-poem-a-day.

CHAPTER 5

‘Out of Season’: The Narrative Ecology of *Persuasion*

Amelia Dale

Austen and the Year Without a Summer

‘I talked of its’ [sic] being bad weather for the Hay–& he returned me the comfort of its’ being much worse for the Wheat’, Austen writes, in a letter to her nephew, James-Edward Austen, on 9 July 1816, recounting a conversation with Edward Woolls.[1] In the same letter she describes how the weather ‘is really too bad, & has been too bad for a long time, much worse than anybody can bear, & I begin to think it will never be fine again. This is a finesse of mine, for I have often observed that if one writes about the Weather, it is generally completely changed before the Letter is read’.[2] Following a sentence which articulates—with the underscored ‘can’ and repeated ‘too bad’—a lament about miserable weather that seems to stretch beyond human endurance, Austen rapidly shifts to a more self-reflexive and optimistic position, reflecting on the changeability and unpredictability of the weather, as well as its capacity for radical localisation: ‘when Mr W. Digweed reaches Steventon tomorrow, he may find You have had a long series of hot, dry weather’.[3] Austen writes with ironic awareness that ‘bad’ weather in Chawton probably spells the same in Steventon, both locations in Hampshire and separated by only a few miles.

A. Dale (✉)
Shanghai University of International Business and Economics, Shanghai, China

A. Collett, O. Murphy (eds.), *Romantic Climates*,
https://doi.org/10.1007/978-3-030-16241-2_5

In 1816 Hampshire, like Europe, was experiencing an unusually cold, wet summer, along with widespread crop shortages. The eruption of Mount Tambora in April 1815 had triggered dramatic climate change. Though no one in Europe at the time connected the volcano to the climate, the eruption resulted in Austen's 'too bad' weather in what was already a cold decade.[4] During this cold, wet summer, Austen was working on *Persuasion*: Cassandra Austen's notes on her sister's composition practices situate her commencement of the manuscript of *Persuasion* in August 1815 and its final completion in August 1816 (though Jocelyn Harris suggests that perhaps she was still revising the manuscript into March 1817).[5] In either case, these dates place *Persuasion*'s composition directly in the wake of the climactic eruption of Mount Tambora in April 1815 and amidst the climate irregularities that followed it.

We can recover hints of historical climatological events, such as Tambora, in Austen's works, via oblique references to events, movements and rhythms that might be 'out of season'. Her work invites reading according to a 'nonreferential aesthetic', to use Mary Poovey's description of how Austen manages the 'historical traces' within her novels, simultaneously registering and deflecting readerly attention from historical events.[6] All Austen's novels contain references to seasons. The thematic centrality of seasonal rhythms is striking, for instance, in *Emma* (1815).[7] *Emma* is also a novel which might be impressed with historical traces of climate irregularities. There has been plenty of attention on Austen's potential 'error' describing apple trees in 'blossom' in June, when Emma decides that for Harriet, Abbey-Mill Farm 'might be safely viewed with all its appendages of prosperity and beauty, its rich pastures, spreading flocks, orchard in blossom, and light column of smoke ascending'.[8] John Sutherland argues that rather than this being a mistake, Austen offers 'a precise depiction, in the form of a miniature montage, of the turning seasons'.[9] A completely different approach is offered by the scientist Euan Nisbet, who in *Nature* subjected *Emma* to a meteorological analysis, taking as the starting point the importance of weather to the novel ('*Emma* is weather') and arguing that the apple trees would have blossomed in the cold weather of 1814 when Austen was writing the novel.[10] The summer of 1814 was unusually cold: the annual mean was one of the lowest recorded in Luke Howard's *The Climate of London* (1833), with the May and June means even colder than those recorded in 1816.[11] *Emma*'s apple trees blossoming out of season therefore have been read as a meticulous description of a climate anomaly and conversely as an agricultural 'montage'. Austenian climates are productive of multiplying opposing readings: as slippages or 'errors', as scientific descriptions or as deliberate aesthetic choices. It is also possible to take into account these multiple opposing readings: climate anomalies can force a year to pass in a moment, in a sentence, in a compression of temporality that might also celebrate rural English life. Climate irregularity—or being 'out of season'—can produce an (un)natural compression of novelistic temporality that becomes part of the novel's narratorial ecosystem.

The same cold summer that may have informed Austen's writing of *Emma* is also the year that *Persuasion* opens. The historical setting of *Persuasion* is therefore bookended by climate anomalies, looking backward and forward to cold summers. Notably precise in its temporality, the novel commences in the cold summer of 1814, after the abdication of Napoleon, which seemed to promise a peace. *Persuasion* is populated by naval officers who have returned from war and are attempting to adjust to this new peace, a peace that would turn out to be short-lived. The novel concludes in February 1815, just before Napoleon's escape from Elba on February 26 prompted a recommencement of war. The Napoleonic Wars finally culminated in the Battle of Waterloo in June 1815. For the first readers of *Persuasion* then, to remember the peace of 1814–1815 is to remember a 'false peace', a brief interlude in the Napoleonic Wars before their more definitive conclusion at Waterloo. *Persuasion* invites readers both to anticipate Waterloo and to contrast their own post-Waterloo perspective with the characters' sense of history. To quote Emily Rohrbach's sensitive analysis: the 'question of how the moment will appear in retrospect, which repeatedly burdens Anne, is one that carries historical weight vis-à-vis the narrator's (and the readers') post-Waterloo perspective on the 1814–15 moment'.[12] Yet not only does *Persuasion*'s temporal retrospectivity demand to be read in relation to global geopolitics but also in relation to what is now understood as the Tamboran climate emergency, itself shadowed by the 'epochal conflict' of the Napoleonic Wars.[13] Given this context, *Persuasion* potentially both looks forward to Waterloo's geopolitical and cultural upheaval and to the climate shifts and the subsistence issues of the 'Year Without a Summer'. In *Persuasion*, where the symbolically fraught seasons are loaded with historical significance, spelling the encroaching recommencement of the Napoleonic Wars, they also, with similarly complex non-linear movements, 'shadow' the climate irregularities of 1816. Climate catastrophe becomes another 'recommencement' following the climate anomalies of 1814. *Persuasion* is a novel of 'anticipated retrospection' as Rohrbach observes, but its temporal rhythms extend to its retrospective anticipation of a disruption of seasonal rhythms.

The Ecology of *Persuasion*

Seasonal variation is thematically and narratologically central to *Persuasion*. The novel tracks carefully historicised seasons of the summer, autumn and winter of the 'false peace' in ways that are intertwined with its concern about the passing of time, political and cultural change and ageing.[14] Take, for instance, the evocative description of Elizabeth's stagnation:

> Thirteen years had seen her mistress of Kellynch Hall, presiding and directing with a self-possession and decision which could never have given the idea of her being younger than she was. For thirteen years had she been doing the honours, and laying down the domestic law at home, and leading the way to the chaise and four, and walking immediately after Lady Russell out of all the drawing-rooms and dining-rooms in the country. Thirteen winters' revolving frosts had seen her opening every ball of credit which a scanty neighbourhood afforded; and thirteen springs shewn their blossoms, as she travelled up to London with her father, for a few weeks annual enjoyment of the great world.[15]

At first glance, the passage offers a dense interplay between a sense of stasis and an inexorable passing of time. The dirge-like repetition of 'thirteen' suggests an entrapped acting out of the same social formalities, with little changing annually apart from Elizabeth's slowly but inevitably ageing body. Elizabeth behaves mechanically and regularly in response to a 'revolving' seasonal cycle. Unlike *Emma*'s apples, Elizabeth's movements are emphatically and fashionably 'in season': she and her father visit London at the height of the London aristocratic 'season' each spring.

Lorri Nandrea argues how repetition with difference, revisiting with variation, animates the structure of *Persuasion*.[16] More than this: it is possible to argue that repetition with difference is inflected with climatological deviations. Specifically, *Persuasion* was written and read with the immediate experience of 'frosts' and 'springs' capable of 'revolv[ing]' in irregular rather than predictable rhythms, and subject to multitudinous variations in length, moments of appearance and severity. *Persuasion*'s seasonal temporal markers therefore potentially denote radical instability and a capacity for a 'revolution … almost beyond expression' (257) in established rhythms that coincide with the novel's striking narratological innovations and attentiveness to socio-political change. Elizabeth and, by extension, the ossified English gentry are attentive to fixed, fashionable 'seasonable' entertainments in a way that contrasts, I will argue, with the 'out of season' travel that Anne undertakes through the course of the novel. Relatedly, the passage, through free indirect discourse, suggests that Elizabeth understands the seasons as smooth, unyielding cycles, 'revolving' about her. This stands in stark opposition to *Persuasion*'s description of the chaotic and fractal structures underpinning seasonal shifts.

How might we begin to read *Persuasion*'s depiction of seasons in light of the climate event that took place in its year of composition? *Persuasion* seems to pre-echo the complex systems that dominate contemporary sciences, particularly the natural sciences. The folding narratological temporality of *Persuasion* needs to be taken account in any historicisation of the novel. Austen's novel is anticipatory, retrospective and consistently nonlinear in its account of events. Rather than just mapping early nineteenth-century English climate catastrophe—and corresponding historical formulations of the climate—onto *Persuasion*, the novel's anticipatory movements invite us to look forward at the way the work might anticipate later models of ecology. This chapter, then, offers a new reading of *Persuasion*, in light of the novel's concern with being 'out of season'. *Persuasion*'s multiple temporalities—its narratorial movements of simultaneous anticipatory retrospection and retrospective anticipation—inflect its treatment of climate. Heidi Scott recently suggests that '[e]cocriticism can theorize how the scientific understanding of nature has literary origins'.[17] To be precise, Scott argues that nineteenth-century 'literary narratives played a seminal role in sketching out the postmodern view of chaotic nature that would emerge in ecological science of the late twentieth century'.[18] Though Scott does not examine Austen, *Persuasion*'s engagement with climate irregularities on the level of narrative design suggests situating it as responding to the climate catastrophe of 1816 with what could be characterised as a poetics of chaos, anticipating contemporary chaotic ecological models.

In arguing that *Persuasion*'s temporality is 'complex', I use the term in its everyday sense, in the etymological sense of 'folded in on itself' and also with reference to 'complexity theory' and theories of non-linear and chaotic systems. N. Katherine Hayles influentially argues that the new sciences of chaos and complexity transform science and the 'epistemic ground' of Western culture, redefining the relationship between order and chaos, spurring a growth of studies on literature and complexity theory.[19] *Persuasion*'s depiction of time, with its 'shaping of a new sense of history' in 'ways that seem magically to run time backward', maps onto its resonant depiction of disruptions and variation in seasonal 'rhythms'.[20] *Persuasion*'s non-linear narratology speaks to chaotic, complex ecological formulations.

Meaning to Have Spring Again

To undertake a *Persuasion*-like return to what has been repeatedly articulated: *Persuasion*'s re-recovery of the past is figured as Anne undergoing a 'second spring', a renewal of her youthful (and crucially, with its blossoming connotations of spring) 'bloom'.[21] Elizabeth's stagnant sense of thirteen winters and thirteen springs 'revolving' lays the groundwork for the multifaceted associative network between the turn of the seasons and Anne's more richly textured consciousness. To give one example: in the walk to Winthrop, when Wentworth speaks earnestly to the younger Louisa Musgrove, Anne resolves that '[h]er *pleasure* in the walk must arise from the exercise and the day, from the view of the last smiles of the year upon the tawny leaves and withered hedges, and from repeating to herself some few of the thousand poetical descriptions extant of autumn' (90). It is possible to detect a slight, sly distance between the melancholy Anne—reader and internal reciter of morose autumnal poetry—and the narrator.[22] Autumn here has its self-consciously conventional figuration as poetic lines perhaps too generic to quote directly in the novel, poetry that articulates 'decline': 'the images of youth and hope, and spring, all gone together' (91). Seasonal imagery is bound up in Anne's making sense of the present in terms of a past lost and a 'youth' that is 'withered'. The passage turns when Anne's walk reaches its climax:

> after another half mile of gradual ascent through large enclosures, where the ploughs at work, and the fresh-made path spoke the farmer, counteracting the sweets of poetical despondence, and meaning to have spring again, they gained the summit of the most considerable hill, which parted Uppercross and Winthrop, and soon commanded a full view of the latter, at the foot of the hill on the other side. (91–92)

Anne's walk has a temporal and affective as well as a spatial dimension. Her movement across the landscape becomes a movement through seasons and their associated affects, the 'gradual ascent' potentially marking a movement away from autumnal despondency. Michael Wood notes the irony here, how 'meaning to have spring again' involves a 'witty mock-confusion of registers' and that there is no reason to presume the farmer has 'poetical despondence'; I would add to this that there is also irony in the way the sentence construction also suggests that the farmer might be ploughing to deliberately counteract Anne's own melancholy musings.[23] This passage is often read as the pastoral practicality of the farmer and his plough countering Anne's Romantic luxuriation in 'the sweets of poetical despondence' (91). The farmer, according to Ashley Tauchert, figures as a practical understanding of the 'cycle of nature'.[24] I want to pick here at the idea that the farmer's plough speaks to a natural, inevitable 'cycle' or at Caroline Franklin's similar observation that 'a second spring is what is natural'.[25] Certainly, and as Tauchert astutely notes, the farmer's plough speaks to a practical, material dependence on the rhythms of the weather. The farmer's labour depends on the seasons, but his desire, his labour, figured in the narration as a 'meaning to have spring' does not *make* the climate. For Austen's contemporary readers, consuming the novel in the wake of the 'Year Without a Summer', a year of catastrophic harvests—'meaning to have spring again' (91)—becomes potentially suffused with precarious longing rather than material certainty.

This registering of the uncertainty of regular seasons is especially poignant given the climactic positioning of the plough in this passage; Anne notices it just before gaining the summit of the hill they are walking up. The plough sows the narratorial seed for the transition from autumn to spring—and spring's variegated associations (bloom, regrowth, youth)—that forward the romantic conclusion of *Persuasion*. The plough, in other words, sets the reader up for the turn (or return) of the narrative's direction at Lyme and Bath, where Anne's 'bloom and freshness of youth' seems to be 'restored' (112). The seasonal shifts fuel *Persuasion*'s simultaneously anticipatory and retrospective motions. The 'revolving' cycle of seasons might be moving forward, time's progress is apparent in the ageing bodies that inhabit *Persuasion*, and its depiction of the aristocracy being supplanted by a meritocratic naval fraternity, but it also wheels in reverse, with the return of Mr. Elliot, the return of Wentworth, the return of romance and Anne's 'bloom'. Yet *Persuasion* resists even this forward/backward linear model: the description of the walk to Winthrop contradicts the sense of a seasonal 'cycle' through jumping between autumn and spring, bypassing winter. The seasonal rhythm is capable of curving, fluctuating, skipping and sticking.

These deviations ripple out from Anne herself; the shifting seasons are sutured to her own perspective and affective state. Seasons and their literary and cultural associations are yoked to Anne's consumption of literature, her consciousness of her age, her social circles and, crucially, her geographic positioning. On an immediate level, Anne's movement between locations through the course of *Persuasion*'s narrative maps onto seasonal transitions (though less fashionably than Elizabeth's movements): Anne leaves Kellynch Hall in late summer, lives in Uppercross in the autumn, visits Lyme as autumn shifts to winter and then finally visits Bath before Christmas. With each relocation comes a renewed awareness that 'a removal from one set of people to another, though at a distance of only three miles, will often include a total change of conversation, opinion, and idea' (45). In Kellynch, in summer, Anne anticipates summer continuing into September in Bath, 'dreading the possible heats of September in all the white glare of Bath', while also looking forward to 'all the influence so sweet and so sad of the autumnal months in the country' (35). A place's local climate in *Persuasion* both actively shapes and is shaped by Anne's own affective relation to it. What is undesirable summer in Bath is 'sweet' autumn in the country near Kellynch.

At the same time, the centrality of Anne's subjectivity foregrounds rather than minimises the way *Persuasion*'s seasonal shifts are catalysed by a network of larger social and cultural forces. When Anne encounters Wentworth, she notes that 'the years which had destroyed her youth and bloom had only given him a more glowing, manly, open look, in no respect lessening his personal advantages' (65). Wentworth (cruelly) resists age, with his 'glowing' look associated with the 'warm' spring-like somatic qualities that Anne recognises as lacking in herself. *Persuasion* quietly chronicles the small window of youthful 'bloom' that marks a woman as sexually desirable, and that a period of years which might render a woman obsolete in the Regency marriage-market may only increase a man's 'personal advantages'. Anne's conversation with Harville at the White Hart further outlines why the bodies of Anne and Wentworth are, at least in the first half of *Persuasion*, marked by different seasons. As well as conceding that male-sexed bodies might be more 'robust' than women's (253), Anne argues that sexed bodies respond to time in different ways: women's bodies are 'longer-lived' and women 'live at home, quiet, confined, and our feelings prey upon us' (253). Men, on the other hand, 'are forced on exertion. You have always a profession, pursuits, business of some sort or other, to take you back into the world immediately, and continual occupation and change soon weaken impressions' (253). Anne's socio-psychological generalisations are emphatically based on her own experience—particularly in the first volume—of undertaking 'many a stroll and many a sigh' upon finding out that Wentworth's sister will be living in Kellynch (32). The climate of *Persuasion* is thus crucially gendered: Anne might have still had her 'bloom' at the commencement of the novel and be less immersed in autumnal affects if she had 'a profession, pursuits, business of some sort' to distract her from her loss of a future with Wentworth (253). So, to observe that *Persuasion*'s tapestry of climatological disruptions is structured by shifts in Anne's mood is not to deny the way that it also speaks to a chaotic, complex ecology. Instead, through Anne's gendered subjectivity, *Persuasion*'s ecology is interimbricated in a network of carefully delineated social and cultural forces: the novel describes an interdependence of 'nature' and 'culture'. Noting the gap between knowledge and action in the twenty-first century with regard to climate change, Joseph Dodds observes that 'it is becoming increasingly clear that what is most needed is *psychological* research, for it is ultimately in human thinking, feeling and behaviour that the problem is generated, and can potentially be solved'.[26] In a different but related sense, *Persuasion* describes climate as inexorably interconnected with thinking, feeling and behaviour.

We also see this through *Persuasion*'s narration of Anne's atemporal 'bloom' being set alongside Sir Walter's complacent commentary on the decaying bodies around him, coloured by his certainty that his own body resists the ravages of time. For example, when he meets his nephew again after ten years, he 'lament[s]' that 'time' seems to have 'altered almost every feature' of Mr. Elliot for the worse. Sir Walter takes Mr. Elliot's compliment 'that he (Sir Walter) was looking exactly as he had done when they last parted' (152) as a statement of fact. Sir Walter's theories of ageing are simultaneously socio-economic and yoked to the climate. He is particularly preoccupied with the way naval officers rapidly age, stating: 'to a degree, I know it is the same with them all: they are all knocked about, and exposed to every climate, and every weather, till they are not fit to be seen' (22). He connects a premature ageing of sailors to the ways in which their travels expose them to 'every climate' and 'every weather'. Not only, then, are sailors literally weather-beaten, with their bodies 'knocked about' by the elements, their travels age them by propelling them through multiple climates. Sailors, moving across the globe, undergo more seasonal transitions in a year than if they had the luxury of 'living on their own property' (23), to the point where their bodies, like out-of-season fashions, are 'not fit to be seen' (22). Sir Walter's aristocratic, non-linear perspective on time, seasons and bodies points again to the way experience of climates in this novel is shaped by both socio-cultural forces and individual subjectivities. Sir Walter's commentary on the decaying, withering bodies about his apparently ageless, aristocratic physique provides a comic counterpoint to *Persuasion*'s focalisation on Anne's complex, embodied experience of time.

Out of Season

Ultimately, *Persuasion*, through its temporal disjunctions, draws on the climate irregularities that shaped its production to explore what it might mean to be 'out of season'. To be 'out of season' might involve having a gendered consciousness of obsolescence, to feel autumn during spring and vice versa, to find that 'to retentive feelings eight years may be little more than nothing' (65), in short to have feelings or a body that has no linear, predictable relationship to time. For instance, when Anne meets Lady Russell to travel to Bath, it is emphatically early winter: 'a dark November day, a small thick rain almost blotting out the very few objects ever to be discerned from the windows' (133). Yet Anne's body seems to resist winter and Lady Russell compliments her, noting Anne's 'improved' looks (134). Just like the farmer's plough anticipating spring in the autumn, there is a quotidian realism to Anne's cheeks turning pink in response to wintery weather. Yet again like the farmer's plough, this seasonal disruption is also couched in joyful anachronism: Anne seems to be 'blessed with a second spring of youth and beauty' (134). Her flesh responds to November's cold weather with spring-like 'bloom'. The movement of the seasons are dramatised, celebrated and, also suggestively—we could even say 'naturally'—disrupted, in unpredictable, chaotic ways.

Dramatic, sometimes ecstatic, moments of being 'out of season' can be identified throughout the book. When Anne and the Musgroves visit Lyme, for instance, their visit is 'entirely out of the season' (107). Visiting a seaside town in early winter, they 'come too late in the year for any amusement or variety which Lyme, as a public place, might offer' (102). The narration continues:

> [T]he rooms were shut up, the lodgers almost all gone, scarcely any family but of the residents left—and, as there is nothing to admire in the buildings themselves, the remarkable situation of the town, the principal street almost hurrying into the water, the walk to the Cobb, skirting round the pleasant little bay, which in the season is animated with bathing machines and company, the Cobb itself, its old wonders and new improvements, with the very beautiful line of cliffs stretching out to the east of the town, are what the stranger's eye will seek; and a very strange stranger it must be, who does not see charms in the immediate environs of Lyme, to make him wish to know it better. (102–103)

The construction of this long sentence is elaborately ironic: the list of the beauties of Lyme follows the dismissal of the buildings themselves, suggesting at first that the qualities might be 'nothing to admire', then, when the semicolon is reached, it is clarified that everything apart from the buildings is worthy of the 'stranger's eye' (102). Though the sentence touches on what Lyme is like 'in season' (102), the passage—and the Lyme section of the novel as a whole—celebrates what a 'deserted and melancholy' (103) Lyme can provide. Contradicting the innkeepers' misgivings about the seasonlessness of their visit, for Anne's company 'nothing was found amiss' (107). Their visit to Lyme might be unfashionably 'out of season', but such temporal disjunction provides its own pleasures, like the architecture of the town itself, where the 'old wonders and new improvements' (102) of the medieval Cobb are enjoyed simultaneously. Despite the shock and disaster of Louisa Musgrove's fall from the Cobb, Anne later recalls the visit with pleasure, exclaiming that Lyme contains '[s]o much novelty and beauty!' (200).[27] To visit a seaside town at the beginning of winter is to travel in a way that patently goes against established movements of genteel, fashionable travel: it is to push against established cultural movements and temporal markers. We can therefore connect the aseasonal, ahistorical travel of Anne's company to Anne's physical recovery of lost time. It is the cold sea breezes of Lyme in early winter that redden Anne's cheeks, which both attracts Mr. Elliot's gaze and leads to Anne apparently seeming, in Wentworth's eyes, to be the woman she was eight years ago, 'something like Anne Elliot again' (112). The travelling to Lyme 'out of season' allows Anne's body to move anachronistically out and against time's forward momentum.

Such temporal and seasonal non-linear narratorial movements are perhaps most overt in the penultimate chapter of *Persuasion*. When Anne and

Wentworth are allowed to have a private conversation and declare their love, there is not only a joyous recapitulation of the past, where they 'exchanged again those feelings and those promises', but a re-revisiting with difference. Anne and Wentworth are not only 'more exquisitely happy' than before, their return to mutual understanding is a re-return, with a formulation that underlines the multiple folds and temporal cross-currents of their time travel: 'There they returned again into the past' (261). The lovers' happiness is crucially anticipatory and forward looking, part of the 'aesthetic of anticipation' Rohrbach identifies in the novel.[28] Together, Anne and Wentworth not only collectively re-recover a future 'projected' (261), that they once thought was lost, this future involves them looking back at this 'present hour' (261). The present becomes coloured by 'all the immortality which the happiest recollections of their own future lives could bestow' (261). To live in the present is to anticipate being viewed retrospectively, as a moment to be recollected in happy futurity.[29]

Anne and Wentworth's romance is also strikingly articulated in terms of contingency, potentiality and probability. Their present happiness is carefully historicised, with Anne and Wentworth spending it absorbed in 'those explanations of what had directly preceded the present moment': all 'the little variations of the last week were gone through; and of yesterday and to-day there could scarcely be an end' (262). Romance not only becomes a process of anticipating the moment as it is seen in the future but also, simultaneously, a review of the events that have led up to it. Yoked to this multifaceted present is the preoccupation with 'describing what never happened', to draw on William Galperin's observations about how the 'missed opportunity' shadows Austen's plots.[30] Yet while Galperin characterises 'what never happened' as 'an alternative history', here at least there is a sense of not just one but infinite 'little variations' evolving into different pasts and presents. Nandrea observes how *Persuasion* articulates multiple and multiplying articulations of what could be possible: 'each present is laden with potential—surrounded by a swarm of "virtual" outcomes'.[31] They rehearse 'the varied potentials of a single moment in time'.[32] Rather than tracing their romantic union to a linear chain of events, for Anne and Wentworth, to examine the past in relation to the present is not only to self-consciously anticipate the future but also to be caught in speculations—infinitely multiplying alternate universes—which resist closure: 'there could scarcely be an end' (262). The events of this climactic day begin with rain, preventing Anne from making a visit to Lady Russell and revealing to her Mr. Elliot's hidden character. Meteorological systems therefore become components in a larger narratorial ecology that brings Anne and Wentworth together: part of the moving network of 'variations' or possible worlds that Anne and Wentworth collectively examine. The weather also shapes the company at the White Hart, with 'Mary and Henrietta, too impatient to wait' leaving 'the moment it had cleared' (249). To speculate about a hidden history of 'missed opportunities' or the infinite 'variations' that Anne and Wentworth's conversation gestures towards is also to be attentive to changes in atmospheric pressure and humidity, to possible worlds where it rained less or more—worlds where Mary and Henrietta remained at the White Hart and the climactic conversation between Anne and Captain Harville did not occur. On another level, such alternative possibilities also speak to the multiple possible decisions that Austen could choose from when plotting Anne and Wentworth's reunion, the 'variations' she worked and reworked in mid-July to early August 1816, revising the sequence of events that would eventually bring Anne and Wentworth together in their simultaneously 'immortal', retrospective and anticipatory present.[33]

It is fitting, given *Persuasion*'s preoccupation with 'being out of season', that this carefully historicised reconciliation between Anne and Wentworth involves them slowly pacing the 'gradual ascent' of a 'comparatively quiet and retired gravel-walk' (261). The description directly echoes the 'gradual ascent' (91) of Anne's autumnal—then spring-like—walk to Winthrop, and that walk's dramatic seasonal skips. When the lovers' walk concludes, the narrator states, 'At last Anne was home again' (266). Romance encompasses a renewal of familiarity, a travelling home, passing through, around and pushing against seasonal transitions. Examining the temporality of Austen's final work, *Sanditon*, Clara Tuite argues that 'to fully engage historicity is to engage anachronicity'.[34] *Persuasion* also demonstrates this, looking anachronistically forward, backward and into other possible worlds. It revisits and reimages moments of disjunctive seasonal movements, when spring directly supplants autumn, and the chills of early winter are embodied as spring. Describing a complex, chaotic system, with weather events refracted through its heroine's networked consciousness, it explores what it might mean to feel, move, live and blossom 'out of season'.

Notes

1. Jane Austen, *Jane Austen's Letters*, ed. Deirdre Le Faye, 4th ed. (Oxford: Oxford University Press, 2011), 330. For another reference to the wet weather in the summer of 1816, see Austen's letter to Anna Lefroy (Chawton, 23 June 1816); she writes of how Mrs. Digweed was caught in 'all the afternoon's rain and was of course wet through', 329.
2. *Jane Austen's Letters*, 329.
3. Ibid., 330.
4. Gillen D'Arcy Wood, *Tambora: The Eruption That Changed the World* (Princeton: Princeton University Press, 2014), 242n1; William K. Klingaman and Nicholas P. Klingaman, *The Year Without Summer: 1816 and the Volcano That Darkened the World and Changed History* (New York: St Martin's Press, 2013), 160–162.
5. Jocelyn Harris, *A Revolution Almost Beyond Expression: Jane Austen's Persuasion* (Newark: University of Delaware Press, 2007), 36–37.
6. Mary Poovey, 'From Politics to Silence: Jane Austen's Nonreferential Aesthetic' in *A Companion to Jane Austen*, eds. Claudia L. Johnson and Clara Tuite (Chichester: Blackwell, 2009), 251–252.
7. Elizabeth Toohey, '*Emma* and the Countryside: Weather and a Place for a Walk', *Persuasions* 21 (1999): 44–52.
8. Jane Austen, *Emma*, eds. Richard Cronin and Dorothy McMillan (Cambridge: Cambridge University Press, 2005), 391. The 'error' was noticed during Austen's lifetime. Her brother Edward joked, 'Jane, I wish you would tell me where you get those apple-trees of yours that come into bloom in July'. Deirdre Le Faye, *Jane Austen: A Family Record*, 2nd ed. (Cambridge: Cambridge University Press, 2004), 230.
9. John Sutherland, *Is Heathcliff a Murderer?: Great Puzzles in 19th-Century Literature* (Oxford: Oxford University Press, 1996), 19; *Who Betrays Elizabeth Bennet?: Further Puzzles in Classic Fiction* (Oxford: Oxford University Press, 1999), 28–33.
10. Euan Nisbet, 'In Retrospect', *Nature* 388.6638 (1997): 137. For a discussion of the possible apples that might be blooming, see Shannon E. Campbell, 'Apples and Apple-blossom Time (Wherein Jane Austen's Reputation for Meticulous Observation Is Vindicated)', *Persuasions: The Jane Austen Journal* 29 (2007): 89–98.
11. Nisbet, 'In Retrospect', 137.
12. Emily Rohrbach, *Modernity's Mist: British Romanticism and the Poetics of Anticipation* (New York: Fordham University Press, 2016), 106. For other critics on *Persuasion*'s complex relation to temporality, see Mary A. Favret, 'Everyday War', *ELH* 72.3 (2005): 605–633; Deidre Shauna Lynch, intro-

duction to *Persuasion* by Jane Austen, ed. James Kinsley (Oxford: Oxford University Press, 2004), vii–xxxiii; William Galperin, *The Historical Austen* (Philadelphia: University of Pennsylvania Press, 2013), 216–238; Lorri G. Nandrea, 'Difference and Repetition in Austen's *Persuasion*', *Studies in the Novel* 39.1 (2007): 48–64; Loraine Fletcher, 'Time and Mourning in *Persuasion*', *Women's Writing* 5.1 (1998): 81–90.

13. Wood, *Tambora*, 5.
14. See Virginia Woolf for an early example of the longstanding critical tradition characterising the novel as 'autumnal'. *The Common Reader: First Series* (1925), ed. Andrew McNeillie (New York and London: Harcourt, 1984), 144. There is a related critical thread connecting *Persuasion* to Shakespeare's *The Winter's Tale* and its description of winter's defeat by spring. Fletcher, 'Time and Mourning in *Persuasion*', 81–90; Harris, *A Revolution Almost Beyond Expression*, 190.
15. Jane Austen, *Persuasion*, eds. Janet Todd and Antje Blank (Cambridge: Cambridge University Press, 2006), 7. All further references to this edition incorporated in the text.
16. Nandrea, 'Difference and Repetition in Austen's *Persuasion*', 48–64.
17. Heidi C. M. Scott, *Chaos and Cosmos: Literary Roots of Modern Ecology in the British Nineteenth Century* (University Park: Pennsylvania State University Press, 2014), 6.
18. Scott, *Chaos and Cosmos*, 3.
19. N. Katherine Hayles, *Chaos Bound: Orderly Disorder in Contemporary Literature and Science* (Ithaca: Cornell University Press, 1990), 16.
20. Rohrbach, *Modernity's Mist*, 110; Lynch, introduction, xii.
21. See, for instance, A. Walton Litz, '*Persuasion*: Forms of Estrangement', in *Jane Austen: Bicentenary Essays*, ed. John Halperin (Cambridge: Cambridge University Press, 1975), 223; Penny Gay, 'The Romanticism of *Persuasion*', *Sydney Studies in English* 5 (1979–1980): 15–30.
22. Michael Wood, 'Time and Her Aunt', in *A Companion to Jane Austen*, eds. Claudia L. Johnson and Clara Tuite (Chichester: Blackwell, 2009), 201.
23. Wood, 'Time and Her Aunt', 201.
24. Ashley Tauchert, *Romancing Jane Austen: Narrative, Realism, and the Possibility of a Happy Ending* (Houndmills: Palgrave Macmillan, 2005), 152.
25. Caroline Franklin, *The Female Romantics: Nineteenth-Century Women Novelists and Byronism* (New York: Routledge, 2013), 96–97.
26. Joseph Dodds, *Psychoanalysis and Ecology at the Edge of Chaos: Complexity Theory, Deleuze|Guattari and Psychoanalysis for a Climate in Crisis* (London: Routledge, 2011), 5.
27. Arguably, Louisa's fall, as part of the machinery that forwards *Persuasion*'s happy ending, helps catalyse Anne's 'second spring'.

28. Rohrbach, *Modernity's Mist*, 109.
29. Ibid., 106.
30. William Galperin, '"Describing What Never Happened": Jane Austen and the History of Missed Opportunities', *ELH* 73.2 (2006): 357.
31. Nandrea, 'Difference and Repetition in Austen's *Persuasion*', 50.
32. Ibid., 58.
33. See Harris, *A Revolution Almost Beyond Expression*, 36–72.
34. Clara Tuite, '*Sanditon*: Austen's pre–post Waterloo', *Textual Practice* 26.4 (2012): 611.

CHAPTER 6

'This Thing of Darkness': Reading Atmospheric Disturbance in Matthew Lewis's *Journal of a West India Proprietor*

Anne Collett

Passing 'the Lizard, the last point of England' on the first of his two voyages to Jamaica, Matthew Lewis remarks on the 'intolerable weather':

> At one in the morning, a violent gust of wind came on ... the waves dashing against the stern, till at last they beat in the quarter galley; the ship too, rolling from side to side ... The captain snuffed out one of the candles, and both being tied to the table, could not relight it with the other ... and thus we were all left in the dark. Then the intolerable noise! The cracking of bulkheads! The sawing of ropes! The screeching of the tiller! The trampling of the sailors! The clattering of the crockery! Every thing above deck and below deck, all in motion at once... 'Confusion worse confounded!'[1]

The confusion of noise and darkness recorded by Lewis in November 1815 is a curious echo of accounts written by witnesses of a volcanic eruption in the Indonesian archipelago in April of that year, and presages the now famous dark and stormy night of August 1816 at the Villa Diodati in Geneva. The 'perpetual tempestuous weather'[2] experienced by Lewis

A. Collett (✉)
English Literatures & Creative Writing, University of Wollongong,
Wollongong, NSW, Australia
e-mail: acollett@uow.edu.au

A. Collett, O. Murphy (eds.), *Romantic Climates*,
https://doi.org/10.1007/978-3-030-16241-2_6

during his voyage to the West Indies and, again, during his time in Switzerland later the following year, might be understood to be causally related to the 1815 eruption of Tambora. But in both instances, the literary record of gothic darkness left to us by members of the Diodati group might be more interestingly understood as the record of personal psychic disturbance that had its roots in social and political unrest (an unrest that both pre- and post-dates the Tambora eruption), but drew its poetics from direct experience of tempestuous weather and prior literary models.

Matthew 'Monk' Lewis is often the forgotten member of the Diodati group. His sojourn in Switzerland with Byron and the Shelleys was brief and, unlike *Frankenstein* or *The Vampyre*, the ghost stories Lewis contributed to the circle have not survived despite his established reputation as a gothic novelist. Lewis had acquired an almost Byronic degree of notoriety after he revealed his authorship of *The Monk* in 1796: Byron himself had mocked him as 'Spectre-mongering Lewis' in 'English Bards and Scotch Reviewers' (1809).[3] After meeting Lewis in late 1813, Byron described him as 'a good and good-humoured man, but pestilently prolix and paradoxical and *personal*'.[4] Lewis was in contact with Byron and their shared circle of acquaintance intermittently from November 1813 through to late 1817, when he departed for his last, fateful voyage to the West Indies.

In the summer of 1816, in the interval between his first and second visits to his Jamaican estates, Lewis joined the Diodati group in Geneva. Lewis, a raconteur of some renown, contributed to the famous storytelling circle with a recitation of the ghost poem he had written for the Princess of Wales and, according to Percy Shelley, a number of 'grim' stories.[5] Lewis had also begun work on a substantial piece of nonfiction that would only be published long after his death: *The Journal of a West India Proprietor* (1834). Penned over the course of his voyages to and residence in Jamaica between November 1815 and May 1818, the *Journal* opens with Byron's fittingly comic-pathetic epigraph: 'I would give many a sugar cane, / Mat. Lewis were alive again!'[6] Evidence of the prolix, paradoxical and personal Lewis—as well as his tendency towards 'grim' imaginings—is everywhere in the *Journal*.

Byron's mocking tribute to Lewis mirrors the nature of Lewis's uneasy authorial position, one that shifts uncomfortably between the genres of abolitionist reportage and the Grand Guignol of gothic melodrama for which Lewis is best known. Throughout the *Journal*, this generic and authorial confusion is signalled by atmospheric disturbance, a jumbling and (in Lewis's word) a 'jangling' of people and ideas. Like the far-reaching consequences of Tambora's eruption, the *Journal* is energised by the unpredictable convergence of events that occur near and far, happenings that are at once distant and yet intimately connected to a disturbance of moral authority.

Over the course of this chapter I want to consider the conflicted nature of Lewis's representation of his role as a 'West India proprietor' as both tangentially and integrally related to the eruption of that volcano in the East Indies. Thousands of leagues of ocean between the islands of Great Britain, Jamaica and Sumbawa are proved a false indicator of distance when the intimate connectedness of family, nation, empire and planet is revealed. The eruption that resulted in the now famous 'darkness' of that Romantic Year Without a Summer might be understood as not so much causing but extending and deepening a darkness that was always already there at the heart of the British imperial 'project'—a project in which Lewis discovers himself to be personally imbricated with the inheritance of his father's Jamaican sugar plantations. The *Journal* is the record of a very complex state of social and personal disturbance that reflects the entanglement of various external and internal pressures—social, political, emotional, philosophical and, unexpectedly, meteorological. These pressures produce a range of climates that interact with each other in sometimes volatile ways. They include a climate of fear induced by the French Revolution and its repercussions; a climate of catastrophism induced by the discoveries and theories of new sciences; a climate of entitlement induced by recent 'successes' of the British imperial project; a climate of guilt induced by religious and philosophical responses to the British slave trade; a climate of literary celebrity induced by the cultivation of Romantic Genius; and a meteorological climate change induced by the eruption of Mount Tambora: all contribute to the curious disturbance of Lewis's autobiographical record of the period.

Matthew Lewis is described by D.L. Macdonald as 'not only the author of *The Monk* (1796), the most sensational of Gothic novels, and of *The Castle Spectre* (1797), the most successful of Gothic melodramas, but also a member of the class that now seems almost mythical but was then taken for granted: the liberal slave-owner'.[7] When his father died in 1812, writes Macdonald, Lewis inherited 'two Jamaican sugar estates, six or seven hundred human beings, and the problem of slavery'. Macdonald remarks that, 'his solution to this problem seems to us ambiguous'.[8] Like many liberal Britons of the period, Lewis was an abolitionist in principle, but not an emancipationist. He supported the abolition of the transatlantic trade in slaves, but not the emancipation of those already enslaved: the men, women and children who, having been converted into property, were labouring in appalling conditions in the Americas. What is interesting here is not so much the ambiguity of his 'solution', but what the record of his two visits to those plantations in 1816 and 1817 reveals of a heart and mind, not so much in a state of ambiguity, as a state of 'jangle': one calmed to some extent by a droll sense of humour and yet also vivified by a Gothic recognition of the human capacity for extreme cruelty and violence. In its ambiguity, misgivings and partial self-awareness, Lewis's *Journal* is a telling document of the broader British imperial venture. It shows his humanitarian impulses to be insufficient, his attempts at reform misguided, half-hearted and ultimately inadequate. In this, Lewis's failed attempt to solve 'the problem of slavery' in the West Indies is matched by similar failings in the East.

Distant Relations

The Year Without a Summer of 1816 proved a grim year for many throughout Europe, but the devastating impact on the Indonesian archipelago is generally absent from north-centric scholarship.[9] This is exemplified even by Thomas Stamford Raffles's *History of Java*.[10] Published in 1817, the narrative of the Tambora eruption on the island of Sumbawa during the lieutenant governor's period of office is merely an addendum. In her *Memoir of the Life and Public Services of Sir Thomas Stamford Raffles* (1859), however, Lady Sophia Raffles reproduces the extracts of reports her husband received during the first days of volcanic activity in April 1815, extracts which offer a more detailed account:

> [*Extract of a letter from Banyuwangi*] At ten, P.M. of the first of April, we heard a noise resembling cannonade, which lasted, at intervals, till nine o'clock next day; it continued at times loud, at others resembling distant thunder; but on the night of the 10th, the explosions became truly tremendous, frequently shaking the earth and sea violently … On the morning of the 3d of April, ashes began to fall like fine snow … On the morning of the 11th, the opposite shore of Bali was completely obscured in a dense cloud, which gradually approached the Java shore, and was dreary and terrific. By one, P.M. candles were necessary; by four, P.M. it was pitch-dark; and so it continued until two o'clock it began to clear up; but the sun was not visible till the 14th, and during this time it was extremely cold … All reports concur

in stating, that so violent and extensive an eruption has not happened within the memory of the oldest inhabitants, nor within tradition.[11]

It seems important to at least remark the effect of what Gillen D'Arcy Wood has described as 'a short-term ecological catastrophe unequalled in the historical record' on the peoples within the immediate vicinity of the volcano: 'Ten thousand peopled died from the immediate explosion and another 100,000 from famine in its aftermath, while tens of thousands became refugees and were sold, or sold themselves, into slavery.'[12] The subtitle of Wood's article, 'Climate, Colonialism, and the Slave Trade in Raffles's *History of Java* (1817)', clarifies a south/north connectedness that does not so much *result* in darkness as it points to an *inheritance* of darkness at the heart of British colonialism and a global slave trade in which Britain was deeply implicated. Wood's brief analysis of Raffles's narrative of colonial governorship during the British interregnum in Java (1811–1816) illuminates the complexity of competing desires and values at play—within the governor himself, within British imperialism, within the interests of world powers and those of the Javanese people.

Raffles understood his role in the terms of a benevolent patriarchy, whose tools were the civilising ones of a liberal politics administered through British law and policy: 'to put down lawless force, to lighten the chain of the slave ... to promote the arts, sciences, and literature, to establish humane institutions'.[13] To this end, he attempted to enforce the Slave Felony Act of 1811 within Java and made numerous raids against the 'pirates' largely responsible for the supply of slaves within the region. Raffles had little success against the sea raiders but a good degree of success within Batavia itself, principally through the dismantling and banning of the slave market. These efforts had, unforeseen, a devastating impact on the most vulnerable in the archipelago. Wood argues that the damage of Tambora's eruption was multiplied exponentially by Raffles's abolitionist stance. His reforms, while having only a limited impact on the 'system of bondage and dependency across the island ... by which Dutch, Chinese, and local chiefs had guaranteed agricultural production for centuries past', nevertheless meant that the survivors of Tambora could not resort to the time-honoured practice of selling themselves into bondage when the alternative was starvation.[14] By undermining a system of patronage, yet neglecting to install modern governmental institutions to replace its 'vital security and relief functions', the British colonial administration in Java 'exacerbated the human impact of the Tambora disaster'.[15]

Rather than understanding Raffles's abolitionist ideals and attempts to implement them as misplaced, Wood would seem here to point to the need to recognise an entrenched and complex social structure that called for a deeper understanding and engagement with the specific natural and social histories of the region. Dismantling the apparatus of a bondage economy required careful planning that recognised the survival mechanisms of slavery, and accounted for them accordingly in the system designed to replace an evil with a good. In addition, the self-portrait offered by Raffles's *History* reveals his conflicted position: whilst identifying slavery as an 'abomination', Raffles argues against immediate and far-reaching abolition. The convolution of his language reveals his recognition of the degree to which pragmatism and/or acknowledgement of the limits of his authority in the region impact upon the implementation of an ideal:

> we could not consistently with those rights of property which were admitted by the laws that we professed to administer, emancipate them at once from servitude, [so] we enacted regulations, as far as we were authorized, to ameliorate their present lot, and lead to their ultimate freedom.[16]

My purpose in bringing to light Raffles's morally uncomfortable position and the rhetorical framework within which it is expressed is to make apparent the similar but different position Lewis occupied as an 'abolitionist slave owner' and the ways in which the language of the *Journal* reveals his discomfort. By positioning my reading of Lewis's Jamaica *Journal* in relation to a discussion of Raffles in Java, I want to signal the connectedness of British imperial ideologies and policies and Romantic-era debates on issues of liberty and equality as these pertain in particular to the slave trade. The impact of these debates is felt throughout Britain's colonies, as much in Java as Jamaica. Wood's analysis of Raffles's character, his role and the complex position of his abolitionist stance, provides a useful model by which to examine the conflicted position and related self-representation of Matthew Lewis, the 'liberal slave-owner' who inherits more than a sugar estate or two in Jamaica.

An Inheritance

When Lewis inherited two sugar plantations in Jamaica on the death of his father in 1812, he inherited an obligation to his family and a responsibility for the large number of enslaved Africans who lived and worked on the plantations. Macdonald observes that 'Lewis gave up literature for sugar': he would devote most of what remained of his short life to managing his plantations. As a literary man in London, Lewis belonged to abolitionist circles agitating for an end to the trade in slaves. Macdonald writes, however, that 'Lewis's attitude towards the institution of slavery was more ambiguous than towards the slave trade. He was not an emancipationist, largely because he thought that emancipation was impossible.'[17] Lewis found himself in a dilemma: the moral position of an abolitionist who had become an owner of enslaved people and someone who directly profited from their unfree labour (as opposed to profiting merely by being a member of a slave-owning family) was surely untenable. How was this to be resolved? Commenting on Lewis's 'attempt to negotiate his discordant roles', Lisa Ann Robertson refers to him as a 'former abolitionist and slave owner', but Lewis certainly didn't consider his position as an abolitionist 'former'.[18]

The abolitionists' opposition was dominated by members of the Society of West India Planters and Merchants. Lewis' father was an anti-abolitionist as well as an anti-emancipationist, as was Lewis's uncle on his mother's side, Robert Sewell.[19] Many of Lewis's connections, however, were drawn from a different cohort. In the British Parliament, the abolitionists were led by William Wilberforce, with whom Lewis became friendly and from whom he sought advice on numerous occasions: particularly upon inheritance of the sugar estates and after his first and subsequent visits to inspect them. Lewis's friends, Lord and Lady Holland, were also abolitionists and 'liberal slave-owners', a position Lewis recognised as problematic. In 1806 Lewis wrote a letter warning Lord Holland of an anti-abolitionist move to make political mileage out of Holland's recent purchase of a large number of 'Negroes' 'for the use of Lady H's Jamaica Estates'.[20] The 'ambiguity' of Lord Holland's position—hypocrisy would be a better word—concerned Lewis because of its potential to discredit the abolitionist's cause. When Lewis inherited his Jamaican estates, he resolved, in contrast, to act upon his avowed abolitionist politics.

Judith Terry claims that Lewis's 'commitment to his humanitarian purpose needs no other proof' than his decision to personally engage in the better management of his plantations not merely once, but twice. Lewis remarks in his *Journal* that 'unless a West-Indian proprietor occasionally visit his estates himself, it is utterly impossible for him to be *certain* that his deputed authority is not abused, however good may be his intentions, and however vigilant his anxiety'.[21] Terry observes that the limits of

Lewis's conceptual framework 'were those of a man of his time' but that he was 'generous, broad-minded, soft-hearted, and engagingly gullible'.[22] This is apparent in the degree to which Lewis acted upon his desire to improve the living and working conditions of his slaves and his surprise at their willingness to take advantage of such 'benevolence'. A journal entry of 4 March 1816 reads:

> Since my arrival in Jamaica, I am not conscious of having omitted any means of satisfying my negroes, and rendering them happy and secure from oppression. I have suffered no person to be punished … I have never refused a favour that I could possibly grant. I have listened patiently to all complaints. I have increased the number of negro holidays, and have given away money and presents of all kinds incessantly. Now for my reward. … (125)

His 'reward' is what he assumes to be feigned illness, the hospital being used as a ruse to avoid work. But the piece is self-mocking with something of a double edge. Lewis, the dramatist, is always aware of the role he is playing, its pretentions and limitations. He nevertheless persists in the work of improvement, or what Robertson refers to as a benevolence that was 'typical of late-stage sensibility': problematically, it is clear that some amelioration of physical brutality and improvement of material conditions not only kept the socio-economic system of slavery intact but was designed to do so.[23] Although Macdonald makes a claim for Lewis's neat exchange of literature for sugar, literature, in the form of a journal—written over the periods of voyage (to and from) and during his sojourn on his Jamaican estates—was the means by which Lewis attempted to juggle the competing demands of his inherited position.

The Text

Lewis embarked on the first of his voyages to visit his Jamaican plantations on 10 November 1815, with the aim of making himself known to his employees and slaves and of acquainting himself with the extent and condition of his recently inherited property. From the outset, the slippage between Lewis's attempts to sympathise with the oppressed and subjugated and his own position of wealth and power is clear, as is his tendency to understand his experiences through a troublingly gothic lens. His journal entry of that date records the 'loudest of all possible shrieks' from a 'despairing pig' who is protesting against boarding the ship via a ladder.

The pig has to be hoisted up in a bucket, taking precedence over Lewis himself, who is hoisted up only after the pig has reached the relative 'safety' of the ship's deck. Here the slave owner paints a comparative portrait of articulated commonality, even 'fellow feeling', between the 'unfree' (the pig) and the 'free' (Lewis). Both 'squeal' against the demands made of them as passengers of the vessel that will imprison one temporarily (and willingly) and the other unwillingly and 'permanently'. One will be borne to safety at journey's end; one is destined to be killed and eaten en route. We could say that although both squeal, the pig indicates real suffering while Lewis indicates only affected suffering—a performance of 'suffering' that is in part a literary device, but might also be read, in light of the representation of relationship between unfree and free that follows, as the luxury of a performance by one on whom no real demands are made and by whom no real claims by others are met beyond the superficial acts of magnanimity and benevolence. Lewis's recognition of the plight of the pig is notable in his first entry of the journal (8 November), in which he describes the misery of waiting for the ship to arrive at Gravesend:

> nowhere else did I see the sky look so dingy, and the river so dirty: to be sure, the place has all the advantages of an English November to assist it in those particulars. Just now, too, a carriage passed my windows, conveying on board a cargo of passengers, who seemed sincerely afflicted at the thoughts of leaving their dear native land! The pigs squeaked, the ducks quacked, and the fowls screamed; and all so dolefully, as clearly to prove, that *theirs* was no dissembled sorrow? And after them (more affecting than all) came a wheelbarrow, with a solitary porker tied in a basket, with his head hanging over on one side, and his legs sticking out on the other, who neither grunted nor moved, nor gave any signs of life, but seemed to be of quite the same opinion as Hannah More's heroine, 'Grief is for little wrongs; despair for mine!'
>
> As Miss O'Neil is to play 'Elwina' for the first time to-morrow; it is a thousand pities that she had not the previous advantage of seeing the speechless despondency of this poor pig; it might have furnished her with some valuable hints, and enabled her to convey more perfectly to the audience the 'expressive silence' of irremediable distress (5).

In this passage, the feeling of 'despair' seems to collapse all distinctions between human and animal, or rather (in a troublingly arch manner that Lewis will repeat throughout the journal), the distinction collapses between pain caused by true suffering and that which is merely performed as a gesture of sensibility. Further complicating and problematising the *Journal's* rhetoric is the way in which non-human livestock (the pig, the ducks) stand in for enslaved human stock. Thus the cargo of 'unfree' passengers who seem 'sincerely affected at thoughts of leaving their dear native land!' is reminiscent of the 'cargo' of enslaved human beings, embarking on the voyage across the Middle Passage to the Americas. Many will not survive the voyage and, of those who do, most will not return to their 'dear native land': these passengers were indeed surely 'sincerely affected'! Lewis's characterisation of the pig's distress makes for clever comedy; but the pig's plight *is* truly 'irremediable' and the animal's distress is accordingly shrill: he is, to all intents and purposes, already 'dead', transformed into a consumable by the appellation, 'porker'. His, like that of the enslaved human beings in the pages that follow, is no dissembled sorrow. But the pig's plight and accompanying shriek, although affecting, is only briefly so, for Lewis is quick to make a play of it: like the suffering of his slaves, it is material for literary embroidery, no more sincere or sympathetic than Charles Lamb's equally arch 'Dissertation on Roast Pig' (1823).

Interestingly, in its final pages, Lewis's journal returns full circle to a pig: not the pig of the voyage upon whom Lewis feasted, but the pigs of the plantation village who are to be 'put to death without mercy' as a fearful example to the hoard of pigs (and their negro owners) who had become a nuisance to the plantation works and grounds of the Great House. Pigs—and enslaved humans—who refuse to be confined to their place must be put in their place. The tyranny of power is performed to real and terrifying effect in the closing lines of Lewis's account of his benevolent management via the instrument of neo-colonialism:

> An order was in consequence issued on a Saturday, that the first four pigs found rambling at large after two days should be put to death without mercy; and accordingly on Monday morning, at the negro breakfast hour, the head governor made his appearance before the house, armed cap-a-pee, with a lance in his hand, and an enormous cutlass by his side. The news of this tremendous apparition spread through the estate like wildfire. Instantly all was in an uproar; the negroes came pouring down from all quarters; in an instant the whole air was rent with noises of all kinds and creatures; men, women, and children shouting and bellowing, geese cackling, dogs barking, turkeys gobbling; and, look where you would, there was a negro running along as fast as he could, and dragging a pig along with him by one of the

> hind legs, while the pigs were all astonishment at this sudden attack, and called upon heaven and earth for commiseration and protection. (252)

The 'whole air rent with noises' recalls Caliban's isle of Shakespeare's *Tempest* (1610–1611), an imaginative model for colonialism that is never far from Lewis's mind. His description of a foiled insurrection highlights the 'jangling' of genres that contribute to the *Journal* and the mentality of the slave owner, with its schlocky gothic horror, its Burkean anxieties about revolutionary activity and its debt to Shakespearean superstition.

Reflecting on the 'fashionable peccadilloes' of the enslaved negroes on his Jamaican plantation in his journal entry of 16 March 1816, Lewis remarks with arch humour that 'unluckily, there are some palates among them which require higher seasoned vices', of which he records, 'poisoning, stabbing, thieving, etc.', and most troubling, a plan 'for giving themselves a grand fete by murdering all the whites on the island'.[24] Lewis then describes a scene—surely embellished by the Gothic imagination of *The Monk*—of a thousand persons engaged in the plot, many of whom were sworn to assist with 'the usual accompanying ceremonies of drinking human blood, eating earth from graves, etc.'[25] 'Luckily', again, the plot is discovered 'in time to prevent any mischief' (137), an odd choice of word, given that of the two captured 'ringleaders', one was sentenced to hanging and the other to transportation. Had the mischief not been curtailed the plot, according to Elizabeth Bohls, 'would have ranked with the largest in the West Indies' long history of slave revolts'.[26] Lewis's use of the word 'mischief' and its association with a (foiled) plot to gain freedom carries the sinister overtones of its use in Shakespeare's *Tempest*, when Caliban urges Stephano to 'Do that good mischief which may make this island / Thine own for ever'—a prompt, in other words, to murder. Lewis's use of the word is almost careless—it appears to undercut the darkness of its usage in much the same way that Stephano and Trinculo's buffoonery undercuts Caliban's very serious attempt to regain sovereignty of his island and his self, by whatever means are available to him.[27] Yet *The Tempest* ends with Prospero's ambiguous ownership of 'this thing of darkness'—a darkness that may refer to the 'poisonous slave' Caliban, or to the 'black magic' of his 'project'.[28] When Lewis implicates Shakespeare's/Prospero's *Tempest* in his own journey into darkness, he situates himself at the heart of a darkness that will also generate Conrad's/Marlow's voyage a little less than a century later; and like the texts that come before and after, his *Journal* is energised by a convergence of events that occur near and far, happenings that are at once distant and yet intimately connected to a disturbance of moral authority.

Lewis's assumption of Prospero's mantle and his attempts to reconcile the role of slave owner with that of liberal, *improving* Georgian gentleman, break down under the pressure of reality in this devil's playground. He reflects that:

> What other negroes may be, I will not pretend to guess; but I am certain that there cannot be more tractable or better disposed persons (take them for all in all) than my negroes of Cornwall. I only wish, that in my future dealings with white persons, whether *in* Jamaica or out of it, I could but meet with half so much gratitude, affection, and good-will. (252)

This gratitude, affection and goodwill are performed under pressure of survival, and are unlikely to be found in less pressured circumstances; but that fine point appears to go undetected by Lewis. His claims must be understood in relation to the attitude of superiority he adopts throughout the journal, that repeatedly assumes negro stupidity where in fact a conscious survival strategy is both more likely and more plausible. But this is not a subject I wish to pursue further in this chapter, rather, I want to consider the turbulence (of all kinds) that punctuates, or rather orchestrates, the drama of Lewis's record and to consider the entangled relationship between meteorological, moral and emotional disturbance.

It could not be claimed with any assurance that the turbulence of the weather at sea and on the island of Jamaica described in Lewis's *Journal* was an accurate record of climate disruption over the years 1815–1818, or that the record of turbulence was directly attributable to the eruption of Tambora. Ships' logs reveal the years 1815 and 1816 to be particularly active Atlantic hurricane seasons, 'with at least 12 tropical cyclones ascribed to 1816 alone', and that 'a northward-displaced Intertropical Convergence Zone appears to be partially responsible for the increased tropical cyclone activity of 1816'—a displacement that might be attributed to atmospheric effects of the Tambora eruption.[29] Letter writers and contributors to a range of British magazines noted the unusually severe winds of 1816 and 1818, 'a great hurricane' effecting northeast Scotland in the winter of 1816, and more surprisingly, severe gales of hurricane strength are recorded in the *Hereford Journal* (where 'hurricanes hardly ever happen'![30]) in January and March of 1818.[31] Lucy Veale and Georgina Endfield record 'numerous localised extreme weather events' in 1816 but observe that '[d]iscriminating between weather effects linked to volcanic events, and the natural variability of the climate is fraught with difficulty', concluding that one 'may conjecture whether these conditions were a function of normal weather, or of volcanic aerosols which would have produced a marked drop in surface pressure across the mid latitudes across the North Atlantic'.[32] This aligns with the suggestion of a 'northward-displaced Intertropical Convergence Zone' that would account for increased hurricane activity in the Caribbean. But rather than theorise a causal link between the eruption and the weather events, I suggest that more is to be gained by reading the tempestuous weather of Lewis's voyage as the representation of another kind of 'convergence zone': the convergence of actual disruptive weather events, state of mind and awareness of a prior literary model, Shakespeare's *Tempest*.

In contrast to the 'noises of all kinds' that rend 'the whole air' of Jamaica, the noises of Shakespeare's isle take the form of 'sweet airs, that give delight and hurt not'. The noises of Lewis's voyage are sounds of tempest that bode no good; they are a cacophony that might be understood to represent the real and the imagined turbulence of the Caribbean from the non-native perspectives of both Lewis and the transported Africans. They might also represent the strain of Lewis's entangled position: the jangle of the conflicted mind/heart of a reluctant slave owner. As indicated, Lewis's journal begins and ends in turmoil; it is a passage through a metaphysical and physical darkness. Lewis's journey into that heart of darkness begins in the 'black melancholy' of the Thames (5). On the second day at sea, Lewis recounts the story of the man who has been with the captain in a violent storm off the Bermudas—the islands upon which Shakespeare's characters are wrecked.

The following day, a tremendous gale blows up, of which the captain says, he has 'never experienced a heavier', but the ship rides the storm successfully. Not so a brig which is 'dashed to pieces completely', and although the crew survive, the pilot 'between bruises, drowning, and fright, had suffered so much, that he died two hours afterwards' (8). The weather continues 'disagreeable and perverse'—veering between violent storm and dead calm—a condition Lewis understands to be expected of these latitudes (13). On entering the tropics, the rains are 'incessant and most violent' (20). On 13 December Lewis describes the capture of a dolphin whose colours—emerald green and gold, silvery grey and azure—'Imagination can[not] picture more beautiful'. 'From the great loss of blood, his colours soon became paler; but when he was at length safely landed on deck, and beating himself to death against the flooring, agony renewed all the lustre of his tints.' When this beautiful creature of the sea is 'quite dead', 'he exhibited nothing but dirty brown and dull dead white' (21).

The following day, the ship enters the latitude of Jamaica, the moon rises against the setting sun. Lewis reports that,

> the clearness of the atmosphere … the smoothness of the ocean, and the aforesaid rose-coloured moon, altogether rendered this sunset the most magical in effect that I ever beheld; and it was with great reluctance that I was called away from admiring it, to ascertain whether the merits of our new acquaintance, the dolphin, extended any further than his kin. Part of him, which was boiled for yesterday's dinner, was rather coarse and dry, and might have been mistaken for indifferent haddock. (22)

The image is again one of disturbance—disturbance of creatures out of their natural element, of violent capture, of equally violent and ugly death, and finally, consumption of one described as 'our new acquaintance'. If the effect attempted is humour, it goes awry. The description is more than faintly horrifying, and again, significantly, capture, death and consumption take place on the deck of a ship in Caribbean waters. This is closely followed by a story equally gruesome, masquerading as 'proof of conjugal attachment' (24). On 16 December, the wind again blows 'so perversely' that the ship is obliged to change course and now steers towards what Lewis describes as 'Shakespeare's "still vexed Bermoothes"' (24). Although Lewis declares himself to be 'greatly disappointed at finding such a scarcity of [the] monsters' he expected to see in tropical seas, the captain soothes his disappointment with a 'true' story of two sharks, frequently seen playing about the ship in Black River Harbour, Jamaica (24). The female is killed by the sailors, leaving the male, apparently, desolate. But the captain recounts how the male shark immediately fell upon his lifeless companion and 'began to eat her up with all possible expedition'. The sailors, 'sensibly excited by so peculiar a mark of posthumous attachment', assist by chopping 'his better half in pieces with their hatchets; while the widower opened his jaws as wide as possible' to gulp down 'pounds and pounds of the dear departed as fast as they were thrown to him' (25).

It is a truly gruesome tale, a ghoulish horror story that one would expect of the author of *The Monk*, but one whose significance 'suffers a sea change' when considered again in the light of slave histories and the desperation caused by the separation people from their homes, of husbands from wives, children from mothers. It recalls the sunset that bloodies the water of Turner's famous painting of the '*Zong* Massacre', in which the 'cargo' of a slaving ship is shown being dumped overboard in the wake of the typhoon, the bodies of enslaved human beings thrown to the sharks for insurance purposes.[33] Avariciousness, violence and death pollute the waters of the Middle Passage, making ripples of disturbance in the mind of one who inherits not only the wealth of the plantation but the darkness at its core.

When Jamaica finally comes into sight on 30 December, the weather is gloomy and the wind and rain are again violent. It is as though the ship itself carries the gloom and the violence, as indeed its forebears like the *Zong* did. On 31 December, a 'black pilot' comes on board, bringing with him an offering—a watermelon, whose pulp is described by Lewis as

> stained here and there with spots of moist red, so that it looks exactly as if the servant in slicing it had cut his finger, and suffered it to bleed over the fruit. Then the seeds, being of dark purple, present the happiest imitation of drops of clotted gore; and altogether ....when I had put a single bit into my mouth, it had such a kind of Shylocky taste of raw flesh about if (not that I recollect having ever eaten a bit of raw flesh itself) that I sent away my plate. (35)

On 1 January 1816, the ship 'squeezes herself into this champagne bottle of a bay' and Lewis joins the celebrations of arrival at last in Jamaica. He steps ashore into a scene of picturesque natural beauty and the charm of exotic pageant. But although the hubbub of noise and movement is 'gay', it is also 'strange' (42).

The sea journey to Savanna-la-Mar, a few miles from the plantation at the western end of the island, is 'lighted by the most beautiful of morning stars' (41) but also blighted by a violent swell that overwhlems Lewis and the crew with sheets of foam. Lewis reaches Savanna-la-Mar and travels by carriage a distance of thirty miles to his estate, where he is greeted by 'uproar and confusion' that 'sets all description to defiance' (41). He remarks that 'whether the pleasure of the negroes was sincere may be doubted; but certainly it was the loudest that I ever witnessed: they all talked together, sang, danced, shouted, and, in the violence of their gesticulations, tumbled over each other, and rolled about upon the ground' (41). Most interesting are the 'sudden bursts of singing and dancing' (42) that are accentuated by the motionlessness of the old women, wrapped in cloaks, who stand at the still centre, their eyes fixed upon the portico Lewis occupies. They put him in mind of the witches of *Macbeth*.

The scene of festivity is marked by an ancient darkness at its core—the witches' gaze is fixed upon him—the slave master, the eater of human flesh. It causes him a moment's uncomfortable pause that is followed later by 'a pang of heart' when the negro house boy introduces himself by saying: 'Massa not know me; me your slave!' (42). Lewis is not so much concerned by the notion of domestic slavery or indeed by slave ownership as he is humiliated by the idea that, 'although the lad appeared all gaiety and good humour, and his whole countenance expressed anxiety to recommend himself to my notice ... the word "slave" seemed to imply, that, although he did feel pleasure then in serving me, if he had detested me he must have served me still' (42). It is at this point that the 'noises of the isle' become a source of irritation rather than one of pleasure; his first night on the island as 'a West India Proprietor' gives him a headache: 'Altogether, they shouted and sang me into a violent headache. It is now one in the morning, and I hear them still shouting and singing' (42–43).

This scene of perturbation, of violent swings from pleasure to misgiving, will be repeated throughout the three months' stay in Jamaica and the return voyage to England of two months in the year 1816. For all the moments of disquiet, however, Lewis does not send away the plate of plantation he is offered, and although he is not enamoured with the flavour, he nevertheless retains the lump of flesh on his pantry shelf. As he leaves the island with little intention of returning, the flesh is left to rot in darkness; a darkness upon which Lewis shines a light briefly and then shuts the door. Does the record of Lewis's travel to and residency on Jamaica bear witness to an initiation into the drinking of human blood and the eating of grave earth that is as much or more his as theirs, as much about self as other? The conflicted nature of the *Journal*, its record of turbulence and the jangle of its telling suggest a representation of a perturbed internal state that is projected onto an external landscape and vice versa. But does this constitute an acknowledgment of 'this thing of darkness ... mine'?

Conclusion

Lewis's *Journal of a West India Proprietor* records the emotional and moral turbulence experienced by a 'reluctant slave owner' during a period of unsettled political and meteorological weather, from late 1815 through to

Lewis's death from yellow fever in mid-1818. Charting the years impacted by the catastrophe of Tambora's eruption, Raffles's *History of Java* reveals the conflicted position of a man similarly caught in the complexity of British colonialism, slavery, Romantic abolitionism, the ideal of benevolent paternal authority, personal ambition and economic imperative. Raffles betrays the difficulty of his position in a language that is either 'too much' or 'too little'. The felt discomfort of Lewis's position is registered in the clash and jangle of noise, colour and smell of the West Indies that assaults his senses and his sensibilities en route and in situ. His is a stormy voyage into darkness that begins (anticipating Conrad) under 'dingy' skies on the 'dirty' Thames where many a slaving ship departed for West Africa, and ends in the limbo of a 'middle passage', Lewis's body being buried at sea to join those others with whom he had more in common than he might have imagined (5).

In the final entry of his journal (2 May 1817) as he prepares to leave Jamaica for the second time, Lewis notes the efforts he has made on the plantation to improve equity and justice. In consultation with his attorney, he devises a code of 'internal regulation' whereby 'the negroes will no longer be punished according to the momentary caprice of their superintendent, but by known and fixed laws ... without respect of partiality or prejudice' (250). Yet he exerts himself whilst 'here' 'to do everything which appeared likely to contribute to their welfare and security during my absence' because he admits that 'in all probability many years will elapse without my making them another visit, if indeed I should ever return at all' (250). The irony here is that Lewis did not, in fact, return: he died of Yellow Fever before reaching England. The superficial changes made in the name of 'equity' are all well and good, but they do not recognise, in fact they refuse to confront, the depth of inequity and, indeed, iniquity of human slavery. Lewis might make a grand statement at the journal's end that '[e]very man of humanity must wish that slavery, even in its best and most mitigated form, had never found a legal sanction', but this is tied to the disclaimer, once again couched in the same language of 'mischief', that 'every man' must also recognise and 'regret' that 'its system is now so incorporated with the welfare of Great Britain as well as of Jamaica, as to make its expiration an absolute impossibility, without the certainty of producing worse mischiefs than the one which we annihilate' (249).

Lewis's inability to come to terms with the ramifications of his inheritance is apparent in his vacillation over his will. After the first trip to his

Jamaican estates, Lewis decided something was needed in the event of his death to ensure the benevolent measures and management style he had implemented would be maintained. His brief visit to the Diodati Villa in the summer of 1816 included the writing of a codicil to his will requiring the future owner of each estate (or son, brother or husband of the owner) to spend three months of every third year on the estate. Failure to do so would mean forfeiture of the property to the next heir. The codicil was witnessed by Byron, Polidori, and Percy Shelley. Lewis's main concern was the future treatment of the men, women, and children whose lives were in his hands. He consulted Wilberforce on many occasions, asking him, just prior to his second voyage, whether it would be advisable to free his slaves in his will; he ultimately decided against such action. After numerous changes to his will, Lewis remained unconvinced that there was any measure he could take to ensure the humane treatment of his slaves after his death and the codicil was revoked. Writing to his mother before embarking on the fateful return voyage of his second visit, Lewis conceded: 'I have done all in my power to secure the poor creatures from further ill-usage; and if my endeavours are marred by others, the crime must rest in *Their* hands, not in mine.'[34] But despite such protestations, or even in light of them, the turmoil of Lewis's journal suggests that he recognised the magnitude of 'the crime' of slavery and the blood left on his hands that would not come out.

Shakespeare's Prospero appears to have offered Lewis a model of princely authority and magical command (through literary arts) of 'his' discovered island and its enslaved inhabitants, but the robes proved to be an uncomfortable fit. The heroic Elizabethan voyages of discovery and associated colonisation are blackened by a Romantic revolution built on the moral authority of liberty and equality. Lewis found himself in an untenable position from which it was difficult to extricate himself. His prayer to the 'Dark Genius' to

> Let not thy strange diseases prey
> On my life; but scare from my couch away
> The yellow Plague's imps (19)

in the early pages of his journal goes unanswered, or perhaps it is a retributive justice meted out by that Dark Genius to those tainted by association with the darkness of slavery, no matter how humanitarian their urges.[35] There is something eerily prescient in his plea; but Lewis 'suffered a sea change', if only in the resurrection of his *Journal*, given new significance by the Abolition Act of 1833. It is unclear why publication of the *Journal* did not closely follow on Lewis's death,[36] but the reason for publication in 1834—the year in which the Slavery Abolition Act took effect throughout the British Empire and Lewis's own slaves were conditionally freed—is clear. Light was shone on its humanitarian impulse and any disruptive darkness was subsumed in the glow of a liberal victory. But, however we understand Lewis's position on slavery, his *Journal* bears witness to a convergence of tempestuous historical forces—political, moral, emotional, literary and meteorological—that produced atmospheric disturbance in his person, his writing and this reader.

Notes

1. Matthew Lewis, *Journal of a West India Proprietor: Kept during a Residence in the Island of Jamaica* [1834] (Oxford: Oxford University Press, 1999), entry dated 'November 19 (Sunday)', 9–10.
2. Lewis, *Journal*, 26 November 1815, 12.
3. Lord Byron, 'English Bards and Scotch Reviewers' (1809) in *Lord Byron: The Complete Poetical Works*, ed. Jerome McGann (Oxford: Oxford University Press, 1980), vol. I, p. 258, l.919.
4. Byron, Journal entry, Tuesday 30 November 1813. *The Works of Lord Byron. Letters and Journals*, Vol. II, ed. Rowland Prothero (London: John Murray, 1898), 356. Original emphasis.
5. Percy Bysshe Shelley, 'Geneva, Sunday, 18th August, 1816' in 'Journal at Geneva', *The Complete Works*, Vol. VI, ed. Roger Ingpen and Walter Peck (New York: Gordian Press, 1965), 147–148.
6. Lewis, *Journal*, frontispiece, n.p. The original version of the verse used as epigraph was included in the 1991 Clarendon edition of Byron's *Complete Works*, to which a note is attached by the editors (Jerome McGann and Barry Weller) pointing out that 'by parodying two lines from Scott's "Lay of the Last Minstrel" (canto v, st.29) Byron adverts to Lewis's major influence on Scott's poetry.' See Terry's Explanatory Notes to the *Journal*, 269.
7. D.L. Macdonald, 'The Isle of Devils: The Jamaican Journal of M.G. Lewis' in *Romanticism and Colonialism: Writing and Empire, 1780–1830*, ed. Tim Fulford and Peter J. Kitson (Cambridge: Cambridge University Press, 1998), 189.
8. Ibid., 189. Discussion of the difference between emancipationists (who believed slaves in the British Empire should be freed) and abolitionists

(who believed the slave trade in the British Empire should be abolished) is discussed in Macdonald's article.

9. David Higgins's book, *British Romanticism, Climate Change, and The Anthropocene* (Palgrave Macmillan/Springer, 2017), is a recent exception. He devotes a chapter to Raffles's account of the Tambora eruption and its effects, describing the account as 'a complex heteroglossic and collaborative production' and 'a material-discursive assemblage in which the natural and the political are thoroughly imbricated' (23). Higgins, *British Romanticism, Climate Change, and The Anthropocene*, 23–53.
10. See Higgins's discussion of the question of unclear authorship of the document that preceded Raffles' *History of Java*, 'Narrative of the Effects of the Eruption from the Tomboro Mountain, in the Island of Sumbawa, On the 11th and 12th April 1815', dated 12 April 1815, and published in the *Transactions of the Batavian Society, of Arts and Sciences* of 1816. The narrative is unsigned but Raffles notes it was 'drawn up by Mr. Assey' although it is also attributed to J.T. Ross, the incoming Society's President. This chapter isn't the place to enter into discussion of debated authorship. Higgins, *British Romanticism, Climate Change, and the Anthropocene*, 28–29.
11. From Stamford Raffles' collection of reports on the impact of the 'extraordinary eruption of the Tomboro Mountain on Sambawa', included in Lady Sophia Raffles, *Memoir of the Life and Public Services of Sir Thomas Stamford Raffles* (1859) (Singapore: Oxford University Press, 1991): 243–244.
12. Gillen D'Arcy Wood, 'The Volcano Lover: Climate, Colonialism, and the Slave Trade in Raffles's *History of Java* (1817)', *Journal for Early Modern Cultural Studies* 8.2 (2008): 35.
13. Stamford Raffles, 'Dedication', *The History of Java*, Vol. I (London: John Murray, 1830), iv.
14. Wood, 'The Volcano Lover', 48.
15. Ibid., 51.
16. Raffles, *History of Java*, Vol. I, 85.
17. D.L. Macdonald, *Monk Lewis: A Critical Biography* (Toronto: University of Toronto Press, 2000), 47–48, 52.
18. Lisa Ann Robertson, '"Sensible" Slavery'. *Prose Studies* 29.2 (2007): 226.
19. See McDonald, *Monk Lewis*, 47–55 for discussion.
20. Letter [exact date uncertain] from Lewis to Lord Holland cited by Macdonald, *Monk Lewis*, 52.
21. Lewis, *Journal* (19 January 1816), 74.
22. Judith Terry, 'Introduction', Lewis. *Journal of a West India Proprietor* (1999): xi.
23. Robertson, Sensible Slavery, 226.

24. Lewis, *Journal*, 137.
25. Matthew Lewis, *The Monk: A Romance* (1796) (Toronto: Broadview, 2004).
26. Elizabeth Bohls, *Slavery and the Politics of Place: Representing the Colonial Caribbean 1770–1883* (Cambridge: Cambridge University Press, 2014), 49.
27. William Shakespeare, *The Tempest* (London: Methuen [Arden Shakespeare], 1976), Act IV, Sc.I, Ll.217–Ll.219.
28. Shakespeare, *Tempest*, Act V, Sc.1, Ll.275–Ll.276.
29. https://en.wikipedia.org/wiki/1810%E2%80%9319_Atlantic_hurricane_seasons, accessed 17/7/2018.
30. This is a reference to the script of *My Fair Lady* based on George Bernard Shaw's *Pygmalion* ('In Hertford, Hereford and Hampshire, hurricanes hardly ever happen!').
31. Lucy Veale and Georgina H Endfield, 'Situating 1816, the "year without summer", in the UK', *The Geographical Journal*, 182.4 (December 2016): 326.
32. Ibid., 326 & 327.
33. J.M.W. Turner, 'Slavers Throwing overboard the Dead and Dying—Typhoon coming on' ('The Slave Ship') 1840.
34. Quoted in Louis Peck's *A Life of Matthew G. Lewis* (Cambridge: Harvard University Press, 1961), 97.
35. Mcdonald notes that Yellow Fever was a disease imported from West Africa into Jamaica with the enslaved Africans, and recounts the results of a survey of British Officers stationed in Jamaica between 1817 and 1836 who died of the disease: 12% per year, 30% during an epidemic; of those who contracted the disease 75% died (*Monk Lewis*, 208).
36. Judith Terry suggests the manuscript was refused by John Murray in 1817 because the payment Lewis demanded was too high. See 'Introduction' to Lewis's *Journal*: x.

CHAPTER 7

When the Earth Moves

Clara Tuite

Introduction

The infamous summer of 1816—the Year Without a Summer—was a summer of brilliant misbegottens, for this was the summer that inaugurated Lord Byron's life of scandalous exile and his catastrophist *oeuvre*. 'Catastrophism pervades the oeuvre in its entirety,'[1] but 1816 is a special and supercharged year that focuses Byron's catastrophist *oeuvre* with the logic of a perfect storm. *Darkness* (1816) takes its cues from that ancient rendition of chaos theory and staple of Enlightenment materialism, Lucretius's *De Rerum Natura*, as well as from Cousin de Grainville's contemporary novel *Le Dernier Homme* (1805), while sullenly-luminously swinging back upon the universe the churlish gaze of this 'rayless' summer of 1816. This poem's apocalyptic fantasy of the extinction of the sun is a prelude to the more sophisticated catastrophism of *Cain, A Mystery* (1821), which explores the future-former worlds then being conjured by the new and newly dissident science of geology, under the aegis of the catastrophist geology of Georges Cuvier; in particular 'the notion of Cuvier, that the world had been destroyed several times before the creation of man' (*CPW* VI, 229). Cuvier appears again, as late as *Don Juan* Canto IX (1823): 'First out of and then back again to Chaos, / The

C. Tuite (✉)
School of Culture and Communication, Faculty of Arts, The University of Melbourne, Parkville, VIC, Australia
e-mail: clarat@unimelb.edu.au

A. Collett, O. Murphy (eds.), *Romantic Climates*,
https://doi.org/10.1007/978-3-030-16241-2_7

Superstratum which will overlay us. / So Cuvier says;' (IX. 37. 295–297), and, indirectly, as early as a late unincorporated addition to *English Bards and Scotch Reviewers* (1809): 'Like reptiles hatched from out the mud of Nile.' *Cain* was itself the belated prequel of *Manfred, A Dramatic Poem* (1817)—'one of the best of my misbegotten' (*BLJ* 5: 249).[2] 1816 is a year of brilliant misbegottens, then, electrifying writing of superb cosmic vision and scientific imagination, which harnesses new theories of the earth and space to explore futures past, the relativity of time, coincidence and chaos, 'Space and eternity—and consciousness' and 'air and earth / Space, and the peopled infinite' (*Manfred*, II.i.47; II.ii.88–89).

Hosting much of the commotion was the Villa Diodati, on the shores of Lake Geneva, where Byron lived from 10 June until 5 October 1816. This is British Romanticism's epicentre of the Year Without a Summer—its 'head-quarters', as Byron wrote to John Cam Hobhouse: 'when you come out—don't go to an Inn … but come on to head-quarters' (*BLJ* 5: 80). As Mary Shelley writes of the June weekend of ghost-story writing that Byron hosted, in her 1831 preface to the third edition of *Frankenstein*, 'it was a wet, ungenial summer, and incessant rain often confined us to the house. … "We will each write a ghost story," said Lord Byron. And his proposition was acceded to.' It was a perfect storm of sociability, sexual tension and creative productivity that used the bad weather of this volcanic winter to conduct experiments in apocalyptic writing. Mary Shelley's *Frankenstein* (1818), John Polidori's *The Vampyre* (1819), and Byron's *Augustus Darvell* (1819, the fragment of a ghost story that later became the foundation of *The Vampyre*), as well as Byron's *Darkness* and *Manfred* and other Diodati poems, such as *Prometheus*, all had their origins here.

My essay traces the early stages of Byron's exile in Switzerland, after the separation in April 1816 from Lady Byron (Annabella Milbanke), to consider how Byron's catastrophist writing mediates emotional disturbance, focusing on *Manfred*, written mainly in September–October 1816 and published in June 1817. Engaging *Manfred* through the emotional apocalypse that Byron was weathering in the wake of the separation, I consider how the storm of scandal and emotional torment surrounding the separation is intimately related to and mediated by the climate of the Year Without a Summer. What is the compositional logic by which environmental geography connects to an emotional geography, at this place and time of Switzerland in the summer of 1816, and beyond?

The essay then moves to the Hampshire countryside and Jane Austen's drafting of *Persuasion*—stalled under the black star of Henry Austen's bankruptcy in the spring of 1816. From here I consider the question of intertextual connections between Austen and Byron, engaging broader questions about literary context and influence, in relation to the rethinking of space and time relations that catastrophism inspires.

Teleconnections

But first, the road—the trail so scandalously blazed—to the Villa Diodati:

> [H]e who is outlawed by general opinion ... whether he be innocent or guilty must undergo all the bitterness of Exile without hope—without pride—without alleviation.— — —
>
> This case was mine. ... I was accused of every monstrous vice by public rumour. ... I was unfit for England ... England was unfit for me— — — —I withdrew—But this was not enough.—In other countries—in Switzerland—in the shadow of the Alps—and by the blue depth of the Lakes I was pursued and breathed upon by the same blight.— — —I crossed the Mountains—but it was the same—so I went little farther, and settled myself by the waves of the Adriatic,—like the Stag at bay who betakes him to the waters.[3]

This 'Stag at bay' might be Frankenstein's creature, but it is Byron going into exile in April 1816, as he narrates it in *Some Observations upon an Article in Blackwood's Edinburgh Magazine* (comp. 1820). This essay—or 'foaming answer in prose'—responds to the malicious 'Remarks on *Don Juan*' that had been published in *Blackwood's* in August 1819.

Here, responding to *Blackwood's* attacks on what the reviewer referred to as Byron's 'selfish' and 'polluted' exile, Byron narrates how he 'retired ... from society—with the exception of one English family' living close to Diodati, which consisted of 'two ladies, a Gentleman—and his son' (*CMP*, 100): Percy Bysshe Shelley, Mary Shelley, Claire Clairmont (Mary Shelley's half-sister) and the Shelleys' son, William. No sooner does Byron move into exile, as London swirled with rumours about the separation, than he becomes the subject of further rumours, that this 'Gentleman [Shelley] and myself were living in promiscuous intercourse with two Sisters, "having formed a league of Incest"' (*CPW*, 100). The rumours were circulated, Byron claims in the essay, by one of the Lake Poets. It was actually Henry Brougham, the Whig lawyer who became Lady Byron's champion after the separation, who spread the rumour, but Byron believed it was the Tory poet laureate, Robert Southey: '"Promiscuous intercourse" could hardly have disgusted the great patron of Pantisocracy—(does Mr Southey remember such a scheme?)—but there was none' (*CMP*, 100–101). So Byron moves from the vortex of one scandal (the separation) into another (the 'league of Incest' with Percy Shelley), retracing his steps to Diodati, as though he were the creature of *Frankenstein*, Mary Shelley's creation born that June weekend. Byron's exile thus anticipates and shadows the exile of Frankenstein's creature, charting his lonely course: 'The desert mountains and dreary glaciers are my refuge. I have wandered here many days; the caves of ice, which I only do not fear, are a dwelling to me, and the only one which man does not grudge.'[4] Shelley's creation recreates the Byron-Shelley domestic setting near Cologny with 'Belrive', the home of the Frankenstein family beside Lake Geneva, thereby honouring Diodati, which was originally named Villa Belle Rive (meaning 'beautiful shore'). And the Villa Diodati in 1816—this Romantic hot spot of time—is thereby traversed in ways that complicate a simple chronology of now and then.

Never published in Byron's lifetime—offered to his publisher, John Murray, then withdrawn, offered again, and finally held back on the advice of Hobhouse and Shelley—*Some Observations* was not published until 1833 (in Thomas Moore's *Works of Lord Byron*). Its narratives of exile, self-vindication and writing back and forward to the future nevertheless anticipate much Romantic writing that appears between 1816 and 1833 and trace intriguing intertextual circuits (from one work to another) and trans-textual circuits (across the life and the work). The tangled intertextual webs and crossings of life and work that inform this extended family and network of Romantics (and create their own 'league of Incest') necessitate ways of rethinking intertextuality and influence that displace chronology. These intertextual and trans-textual traversals disturb chronology and demand not just a 'polytemporal view of textual meaning',[5] but a properly preposterous view of textual meaning. Such preposterous temporalities and traversals across writings and authorial lives are the very stuff of Romanticism. They embody too the principle of deferred or distant relations between cause and effect that, in climate and ecological science, goes by the name of 'teleconnection' and which explains the principle of climate variability whereby the distant effects of climate events in another place are understood not as random but as related.[6] In *Tambora: The Eruption That Changed the World*, Gillen D'Arcy Wood defines teleconnection as 'the physically remote relation between cause and effect' and analyses 'the Tambora climate emergency of 1815–1818'—which had a three-year impact of effects such as persistent rain, low temperatures, poor harvests, food riots, and the Year Without a Summer—as a 'case study' in teleconnection.[7] 'Volcanic strife traveled great distances and via obscure agents. But it is only by tracing such "teleconnections"—a guiding principle of today's climate and ecological sciences—that the *worldwide* tragedy of Tambora can be rescued from its two-century oblivion.'[8] The reason for that oblivion? Napoleon.[9]

Taking my cues from Wood's use of 'teleconnections' to open up a dialogue between literature and climate science, I adopt the idea of 'teleconnections' as a way of rethinking space-time relations, and with it questions of literary-historical periodization, intertextuality and influence, as well as how Romantic literature mediates geological and emotional disturbance. I engage a fairly focussed intertextual archive that includes *Manfred* and its critical reception, Byron's letters of 1816–1817 from the Villa Diodati and after, as well as the *Alpine Journal* of 17–29 September 1816 that Byron wrote for his half-sister Augusta Leigh while he toured the Bernese Oberland with Hobhouse ('To you—dearest Augusta—I send—and *for* you—I have kept this record of what I have seen & felt.—Love me as you are beloved by me.— —' [*BLJ* 5: 105]). The *Alpine Journal* has always been a vital intertext for *Manfred*, and indeed was used by Moore to footnote the text of *Manfred* in the *Works* of 1832.

This intertextual archive attests to the emergence of what Noah Heringman, emphasizing the 'literariness' of geology, refers to as a 'shared vernacular' of landscape aesthetics between Romanticism and geology.[10] Essentially, this is a linguistic paradigm of catastrophist environmental discourse that mediates emotional, social and aesthetic effects, and is applied simultaneously to natural environments, literary works, social figures and emotional experiences. So strong is this sense of a shared paradigm for Ralph O'Connor that he reads 'science *as* literature'.[11] Catastrophism spans a wide range of contemporary geological positions, from the conservative view of catastrophe as an effect of divine wrath to the more gradualist views developed in James Hutton's *Theory of the Earth* (1795) and apotheosized in Charles Lyell's *Principles of Geology* (1830).[12] Wood notes that it is only with 'the advantage of modern scientific instruments and data [that one can] "see" the otherwise invisible teleconnections linking tropical eruptions, climate change, and human affairs'.[13] Nevertheless, these mysterious connections (if not actually seen) certainly make their presence keenly felt in Romantic geological discourse. As Lyell writes in *Principles of Geology* (1830), 'The form of a coast, the configuration of the interior of a country, the existence and extent of lakes, valleys, and mountains, can often be traced to the former prevalence of earthquakes and volcanoes, in regions which have long been undisturbed. To these remote convulsions the present fertility of some districts, the sterile character of others ... the climate, and various particularities, may be distinctly referred.'[14] Hence, the sense of the mystery of deep time and distant causes informed both geological thinking and Romantic poetry alike. And the influence worked both ways. As O'Connor compellingly argues, Byron's poetry 'both prompted and helped geologists in Britain promote their science,' it 'stimulated, sustained, and popularized the science', and was quoted extensively by geologists as part of this process.[15]

As I suggest, in Byron's *oeuvre*, catastrophism works as a form of magical historical thinking that engages the natural world through a conjunction of the sublime and the toxic. *Manfred* registers catastrophism in its status as both spectacular aestheticization of violent natural change and as a critical and metaphysical enquiry into the history of nature and natural change, at the same time as it mediates these geological debates.

Teleconnections refer to the effects of an action in one place being felt in another more distant place, precisely what happens when the earth moves through the effects of a volcano or earthquake. Byron's catastrophist writing combines metaphysical questioning and scientific imagination to explore the relationship between the seen and the unseen or what is felt but not seen. And what better way to explore the effects of an action in one place being felt in another than through 'a sort of mad Drama' (*BLJ* 5: 188), about 'a kind of magician' (*BLJ* 5: 170) who busies himself with 'Mysterious agency!' (I. i. 28), allegorizing Byron's literary agency as a kind of magic. Manfred the magician and metaphysical questioner also channels some heavy emotional weather: 'tormented by a species of remorse—the cause of which is left half unexplained', as Byron writes to Murray, when he first mentions the drama in February 1817 (*BLJ* 5: 170).

This withholding of causality was registered (censoriously) in the drama's critical reception. The *Eclectic Review* attacks *Manfred* as 'without plot and without purpose', while John Wilson in *Blackwood*'s attributes 'imperfection, incompleteness, and confusion ... to the inherent mystery of the subject', while the *European Magazine* complained that Byron 'has given us effects without detailing causes, and shown us strange things without a clue to their development.'[16] Enquiring into causality—the origins of the earth—was a key preoccupation of geology; and the enquiry into remote and unseen causes was a source of intrigue and sublimity. As Horova notes, 'Scientific treatises, the new geological gospels, uncovered the miracles of theretofore unfathomed and unfathomable depths of time.'[17] But geology's capacity to challenge divine causality also rendered it suspect to many as an infidel science.[18] Byron transfers this scientific enquiry to the world of the emotions, when Manfred conceives a journey to the core of his heart: 'As I approach the core of my heart's grief.' In the world of the emotions, the site of effects without causes, of torments left unexplained—what can be felt but not seen—is of course the unconscious. A machine of projection, casting memories of the past into the future ('My pang shall find a voice', II.ii.50), while as much a device of remembering, the unconscious also manufactures 'Forgetfulness' (I.i.136) and 'Oblivion, self-oblivion' (I.i.144), oscillating between the two.

In the Smoky Distance

The cause of Manfred's remorse is left unexplained because it is unspeakable. An iconic Byronic brooding hero, Manfred broods upon a guilty passion for his half-sister Astarte. So the cause of Manfred's remorse was arguably the unnameable sin of incest, which was nevertheless promptly named in most contemporary reviews of the work, which was read at the time—and has been since—as evidence confirming the rumours of Byron's incestuous relations with his half-sister Augusta. Many reviews, like *The Gentleman's Magazine*, simply denounced Byron for daring to represent such an unspeakable crime: 'Manfred has exiled himself from society; and what is to be the ground of our compassion for the exile? Simply the commission of one of the most revolting of crimes. He has committed incest!'[19]

Francis Jeffrey, in the Whiggish *Edinburgh Review*, praised the poem for the skill with which it represented incest under the sign of an appropriate obscurity. In that sense, he was unusual among contemporary reviewers for recognizing the absence of a cause as part of the play's dramaturgical design. A generally sympathetic and astute contemporary reader of Byron (if also a by no means uncritical and often ambivalent one), Jeffrey at first registers the representation of incest as 'painful and offensive', but then concedes that the poem's 'dim and magnificent sketch of a subject which did not admit of more accurate drawing, or more brilliant coloring' is itself an achievement. The representation of incest is striking, then, precisely because it is obscure, and therefore to some extent appropriate (and not diabolical):

> It all springs from the disappointment or fatal issue of an incestuous passion; and incest, according to our modern ideas—for it was otherwise in antiquity—is not a thing to be at all brought before the imagination. …
>
> Its obscurity is a part of its grandeur; and the darkness that rests upon it, and the smoky distance in which it is lost, are all devices to increase its majesty, to stimulate our curiosity, and to impress us with deeper awe.[20]

Manfred, then, is like a volcano itself: brooding in 'the smoky distance', obscure, sublime, with darkness resting on it. This is the literary work as natural catastrophe or revelation (as it was for its author, with the pervasive contemporary figuration of Byron as a comet or meteor).

As well as elaborating obscure, unspeakable human emotions and the unconscious, the play is also captivated with the sublimity of space—the domain of the nonhuman. So much so that William Roberts in the *British Review*, extending the general complaint about the lack of development, observed that 'One only character has absorbed … Lord Byron's creative power: "The steady aspect of one clear large star," of demoniac influence, has fascinated his genius, and we perfectly despair of ever seeing the spell broken.'[21] Indeed, the Byronic hero is himself pantomimically subordinated ('Thou worm!' I.i.125) to this star, a succession of spirits, and to the maternally appropriating 'space' that births the star: 'Space bosom'd not a lovelier star' (I.i.115). Hence another reason contemporary readers were troubled by the play: this turn away from 'character' to the earth and stars and outer space was regarded by some as a form of godlessness (just as geology itself was regarded as an infidel science). *Manfred*'s rhapsodic conjuration of the cosmological sublime is imbricated therefore with both the romance of geology and the threat this romance posed to orthodox Christian beliefs.

Jeffrey, on the other hand, was highly taken with the particular conjunction of character and scenery in *Manfred*, praising the 'still and delicious witchery' that marked Act II, scene II and the 'celestial beauty of the being who reveals herself in the midst of these visible enchantments.'[22] This 'being' is the Witch of the Alps, one of the many spirits (or 'wild Personifications,' as Wilson referred to them in his review in *Blackwood*'s) conjured by Manfred: 'Beautiful Spirit! With thy hair of light.' She appears in 'A lower valley in the Alps—A Cataract', which Byron casts as a sublimely apocalyptic spectacle:

> It is not noon—the sunbow's rays still arch
> The torrent with the many hues of heaven,
> And roll the sheeted silver's waving column
> O'er the crag's headlong perpendicular,
> And fling its lines of foaming light along,
> And to and fro, like the pale courser's tail,
> The Giant steed, to be bestrode by Death,
> As told in the Apocalypse. No eyes
> But mine now drink this sight of loveliness; (*M* II.ii.1–9)

The 'delicious witchery' involves a complex tonality, as the gorgeous evocation of sun and 'foaming light' morphs into the Gothic figures of 'The Giant steed' and 'Death', quickly followed by the bathetic and slightly schlocky, campy Gothic figure of 'No eyes'. This Gothic effect is created by the enjambment whereby 'No eyes' seems to belong (or not) to the apocalyptic horse, as the poetic line hangs with this suggestion—as though pausing to look over the precipice—in a kind of arrested enjambment, before running over into the next line to be connected with the first-person 'mine': 'No eyes / But mine.' Somewhat comically, the eyes are reunited in the next line with the poem's speaker, Manfred, and a vision of Death is thus averted by the vision of 'loveliness'. With such spectacular enjambment (whose prerogative is overflow—appropriately for this stanza on the gushing waterfall), Byron too, like the 'sunbow', 'flings his lines' 'headlong'.[23]

Not just a vision of loveliness, the Witch is canny, too, coaxing Manfred into sharing his secret. In a moment of coy if lugubrious understatement, Manfred, the loner, conversing with skulls, tells the Witch he

> … dived,
> … to the caves of death,
> Searching its cause in its effect; and drew
> From wither'd bones, and skulls, and heap'd up dust,
> Conclusions most forbidden. (*M* II.ii.79–83)

This performative referring to while refusing to name 'Conclusions most forbidden' draws attention to the scandalous rumours that were swirling about in the wake of the separation, in September–October 1816, when

Byron began working on *Manfred*, and were still swirling in June 1817, when the drama was published. What are these 'Conclusions'? The drama puts it out there and then withholds. But it then goes on to broach the forbidden topic, and to speak the unspeakable (Manfred's forbidden love for Astarte which relays Byron's forbidden love of Augusta), in those beautiful heavily dash-laden moments that dramatize Manfred hesitating to name, with the Witch urging him on: 'Proceed' (II.ii.97), and Manfred then acceding to proceed:

> As I approach the core of my heart's grief—
> But to my task. I have not named to thee
> Father or mother, mistress, friend, or being,
> With whom I wore the chain of human ties;
> If I had such, they seem'd not such to me—
> Yet there was one—
> WITCH Spare not thyself—proceed. (II.ii.99–104) …
> MAN I loved her, and destroy'd her!
> WITCH With thy hand? (*M* II.ii.117)

Manfred's journey to his heart's 'core' conjoins emotional with geological discovery, as emotional investigation is figured like a journey to the earth's core.

Spectral Projections

Spectral presences are important here too, as materializations of different kinds of speech acts achieved through poetic practice: projections, conjurations, externalizations.[24] Spectral, but relaying the here and now, as *Manfred, A Dramatic Poem* negotiates Byron's new position of authorial exile, imagined as that of a magician: 'Now to my task.—Mysterious agency!' (*M* I. i. 28). The figure of the magician mediates Byron's fascination and anxiety, at this early stage of exile, with the idea of being in one place (Switzerland, then Italy) while being able to produce effects in another (England). *Manfred*'s performative dramas of magical agency and projection both farcically dramatize and enact the challenge of maintaining a reading audience in England from a position of exile in Europe; they mediate fantasies of authorial control in a context of radically attenuated relations between author and audience; and they stage the task of addressing an audience as the work of making teleconnections, with readers remote not only in place but also in time.

That Manfred the magician is a figure for Byron the author, Byron takes some pleasure in reminding his readers in *Some Observations* (at Southey's expense): 'I never heard that … notwithstanding all the magicians in Thalaba—any body has ever taken Mr Southey for a Conjuror. — — — Whereas I have had some difficulty in extricating me even from Manfred—who as Mr. Southey slily observes … "Met the devil on the Jungfrau—and bullied him"—and I answer Mr. Southey—who has apparently in his political life not been so successful against the great Enemy, —that in this Manfred exactly followed the sacred precept— "Resist the Devil and he will flee from you"' (*CMP*, 90). The question of the devil is a live one for geology and catastrophism. Here, in *Some Observations* (1820) Byron mimics the parodically pious preface of *Cain* that had taunted literalist believers and pious scandal-mongers (like Southey) with the charged figure of Cuvier, and shown how hard liberal Anglican divines had to work to make geology and catastrophism religiously palatable.[25]

Inverting space-time relations, Byron's catastrophism also involves—as well it might—an inversion of the conventional climates of heaven and hell. In the *Alpine Journal*, Byron revisits 'Avalanches falling every five minutes nearly—as if God was pelting the Devil down from Heaven with snow balls' and 'like the foam of the Ocean of Hell during a Springtide—it was white & sulphery', and notes how 'on arriving at the summit we looked down the other side upon a boiling sea of cloud' (*BLJ* 5: 102). Such figures as 'Ocean of Hell' and 'boiling sea of cloud' suggest a form of what Timothy Morton calls 'hyperobjects', or 'objects so distributed across space and time as to transcend spatiotemporal specificity'.[26]

Such objects, such nonhuman actors of the natural world are self-conscious presences, like 'the clouds' that 'Pause to repose themselves', as the First Destiny explains:

> And this most steep fantastic pinnacle,
> The fretwork of some earthquake—where the clouds
> Pause to repose themselves in passing by—
> Is sacred to our revels, or our vigils; (*M* II, iii, 9–12)

This figure of the clouds—at once stretched out and bounded by Byron's generous dashes—performs a certain self-reflexivity: the clouds don't stop to rest, but 'pause to repose themselves'. What does it mean, exactly, this lovely archaism with its slightly excessive transitivity? For the figure is only passingly about resting or reposing; it conjoins reposing while passing by, reposing *in* passing by. This is a marvellous figuration of the scientific phenomenon by which clouds seem to rest as they move—as though deliberately arranging themselves as part of this 'fantastic' aesthetic spectacle, in order to be framed by the 'fretwork of some earthquake'—just as the dashes are the fretwork for their typographical framing on the printed page. The pinnacle that enables this spectacle is 'sacred', 'to our revels, or our vigils', where the 'sacred' designates an imbricated emotional, aesthetic and spiritual realm.

The 'fantastic pinnacle' also represents sublime misanthropy and exile, for the social monster, as Manfred confides in the Witch:

> I said, with men, and with the thoughts of men,
> I held but slight communion; but instead,
> My joy was in the Wilderness, to breathe
> The difficult air of the iced mountain's top,
> Where the birds dare not build, nor insect's wing
> Flit o'er the herbless granite; (*M* II, ii, 60–65)

Belying the sense of emptiness it conjures, the figure of the 'herbless granite' brings a rich hinterland of intertextual associations from the Diodati writings of 1816. *Herbless* repeats one of the negative adjectival keywords of *Darkness*, with its dirge-like procession of words ending in 'less' ('Rayless, and pathless', 'moonless', 'Seasonless, herbless, treeless, manless, lifeless' (line 71), 'useless', 'stingless', 'tombless', 'sailorless'). So too is this form used in *Augustus Darvell* (comp. 1816), whose final word presents an intriguing figure of emotional apocalypse: 'between astonishment and grief I was tearless' (*CMP*, 63). This form occurs again in the *Alpine Journal*: 'Passed *whole woods of withered pines—all withered*—trunks stripped & barkless—branches lifeless—done by a single winter—their appearance reminded me of me & my family.—' (*BLJ* 5: 102). Here, in a poignant moment of bare anthropomorphism, emotional and geological catastrophe are conjoined.

As for the granite, 'no rock is grander in form nor more sublime in structure', announced Humphry Davy, the celebrity chemist and scientific lecturer.[27] The reference to granite indexes vital catastrophist debates. Were rocks such as granite made by fire, as James Hutton and his supporters (the Plutonists) argued, or were they formed by water, as deposits from a universal ocean, as Abraham Werner and his supporters, such as Robert Jameson at Edinburgh (the Neptunists), claimed?[28]

Moving Granite

This sublime and obdurate stone could also be 'pretty', as it was in 1816, when the earth moved through the souvenir gifting of granite stones from Mont Blanc that Byron makes for Augusta and his daughter Ada. As he writes to Augusta:

> I am going to Chamouni (to leave my card with Mont Blanc) and I mean to buy some pretty granite & spar playthings for children (which abound there) for my daughter—and my nieces—you will forward what I select to little Da—& divide the rest among your own. I shall send them by Scrope. (Diodati, Aug 27, 1816 *BLJ* 5: 89)

Byron is very taken with the gifts (and in touristic fashion, the souvenirs themselves were one of his reasons for travelling—'I mean to buy'). After Byron's return from Mont Blanc, a number of letters (mainly to Augusta but also to Annabella) remind their addressees of the impending arrival of these gifts, en route with Scrope Berdmore Davies. Attentively, even obsessively, Byron negotiates anxieties about having to deliver the gifts indirectly: 'By Mr Davies I sent you for yourself—little Da—& my nieces, a variety of Chrystal & other trinkets from Mont Blanc & Chamouni, which I got upon the spot for you all. I hope they will arrive safely' (Sept 14, *BLJ* 5: 93). And it was the same with Byron's manuscripts, sent from exile to Murray, in London, via various intermediaries, such as Scrope, Hobhouse and Polidori.

As Byron writes to Augusta, from Diodati:

> [Scrope] is charged with a few packets of seals—necklaces—balls &c.—& I know not what—formed of Chrystals—Agates—and other stones—*all of & from Mont Blanc* bought & brought by me on & from the spot—expressly for you to divide among yourself and the children—including also your niece Ada, for whom I selected a ball (of Granite—a soft substance by the way—but the only one there) wherewithal to roll & play—when she is old enough—and mischievous enough—and moreover a Chrystal necklace—and anything else you may like to add for her—the Love!— —The rest are for you. … I hope Scrope will carry them all safely—as he promised— —

> There are seals and all kinds of fooleries—pray—like them—for they come from a very curious place (nothing like it hardly in all I ever saw)—to say nothing of the giver. (Sept 8, 1816, *BLJ* 5: 91)

Again, the granite proved to be intriguingly changeable, indeed tractable: 'a soft substance by the way'. And Byron delights in the stones' capacity as children's toys ('all kinds of fooleries'). The stones suggest a form of souveniring and gifting as a mode not only of remembering but also of being remembered, which was vitally important to Byron at this early moment in exile. Byron's appeal to Augusta to 'pray—like them—' (as though implicitly to 'like' 'the giver') registers with pathos the vulnerability of being newly in exile, a pathos underscored by the final appeal with which Byron signs off the letter (like a teenage girl): 'Don't hate me' (*BLJ* 5: 92).

Byron gifts his loved ones back in England pieces of the earth from his place of exile, as though to secure connection in spite of distance, clinging to the materiality of the here and now while conjuring a drama about spectral projection. Just as the souvenir stone is a synecdoche of place ('a very curious place'), so too is it a synecdochic logic by which Byron's souveniring of pieces of granite from Mont Blanc ('on & from the spot') connects geological disturbance to the emotional disturbance in Byron's life.

For in the letters to Augusta about the souvenirs, Byron also writes about the separation, conjoining a storm of scandal with a meteorological storm, as he works through the separation, and projects 'the Alps' as his place of refuge:

> While the storm lasted & you were all pressing & comforting me with condemnation in Piccadilly—it was bad enough—& violent enough—but it is worse now.—I have neither strength nor spirits—nor inclination to carry me through anything which will clear my brain or lighten my heart.—I mean to cross the Alps at the end of this month—and go—God knows where—by Dalmatia—up to the Arnauts again—if nothing better can be done;—I still have a world before me—this—or the next. (Diodati, Sept 8, 1816, *BLJ* 5: 92)

Byron crosses the Alps, and re-crosses them through Manfred, whose metaphysical discussions of space and time are freighted with the bitter emotion of heartbreak. As Manfred says to the Chamois Hunter (the poem's Wordsworthian stooge)[29]:

Think'st thou existence doth depend on time?
It doth; but actions are our epochs: mine
Have made my days and nights imperishable,
Endless, and all alike, as sands on the shore,
Innumerable atoms, and one desart,
Barren and cold, on which the wild waves break,
But nothing rests, save carcasses and wrecks,
Rocks, and the salt-surf weeds of bitterness. (*M* II, i, 51–58)

Wrecks is a special Byronic word, charged with a particular resonance throughout the *oeuvre* reserved for broken hearts. Here it is given a Neptunist inflection, set in apposition with the 'Rocks, and the salt-surf weeds of bitterness'. In this same letter to Augusta of 8 September, Byron writes about the separation from Lady Byron: 'I had enough before I ever knew her … but this last wreck has affected me very differently' (*BLJ* 5: 91). One of the remarkable things about Byron's post-separation, exilic catastrophist poetry is how its cosmological speculation about future-former worlds cohabits with the personal, and the marvellous force and resonance the word *wreck* takes on here, a word meaning both worlds and hearts. In Byron's world, the wreck of a broken heart is given the same existential force and weight as that of a long-lost planet. The reference to 'wrecks' suggests a kind of catastrophism as stoicism that Byron brings to his philosophical reflections on love.

A Heart-Quake

Manfred revolves around destructive forms of love, as it mediates the separation and its broken hearts, the memory of lost love, the end of love. However, Byron's catastrophist *oeuvre* is also fascinated by the equally apocalyptic moment and sensation of falling *in* love—the thunderclap moment. (Call it the Carole King moment: 'I feel the earth move under my feet / I feel the sky tum-b-ling down, I feel my heart start to trem-b-ling.' And King's song, celebrated for its mellow confessionality, reached its own June summer peak as no. 1 on the *Billboard* Hot 100 on 19 June 1971.) This earth-moving sensation is registered as the 'heart-quake' moment of *Don Juan*, Canto II (1819), the emotional heavy-weather event that occurs in the early days in a relationship, when for Juan and Haidee:

Their lips drew near, and clung into a kiss;

A long, long kiss, a kiss of youth, and love,
And beauty, all concentrating like rays
Into one focus, kindled from above;
Such kisses as belong to early days,
Where heart, and soul, and sense, in concert move,
And the blood's lava, and the pulse a blaze,
Each kiss a heart-quake,—for a kiss's strength,
I think, it must be reckon'd by its length.

By length I mean duration; theirs endured
Heaven knows how long—no doubt they never reckon'd;
And if they had, they could not have secured
The sum of their sensations to a second;
They had not spoken; but they felt allured,
As if their souls and lips each other beckon'd,
Which, being join'd, like swarming bees they clung—
Their hearts the flowers from whence the honey sprung. (*Don Juan*, Canto II, stanzas 185–7, lines 1480–96)

Byron uses this happily apocalyptic moment of falling in love to explore the metaphysical concepts of time and duration: the temporality of love. And the catastrophist figures of the volcano and earthquake are used to figure love ('the blood's lava' and 'Each kiss a heart-quake'). The poem dramatizes love's capacity to transform perceptions of time, as in the witty insight about measuring the length of a kiss. (If Juan and Haidee had been counting, they would never have lost themselves enough to be *in* their sensation; their love would not have lasted a second. Instead, it 'endured', as Diana Ross and the Supremes sang in 1968: 'Forever Came Today'.)

Another metaphysical reflection about the endurance of love is delivered by Donna Julia, to Juan, in the wash-up of their forbidden affair, after Juan is sent abroad and Julia to a convent, where she writes to him:

'I loved, I love you, for that love have lost
State, station, heaven, mankind's, my own esteem,
And yet can not regret what it hath cost,
So dear is still the memory of that dream;
Yet, if I name my guilt, 'tis not to boast,

None can deem harshlier of me than I deem:
I trace this scrawl because I cannot rest—
I've nothing to reproach, nor to request.

'Man's love is of his life a thing apart,
'Tis woman's whole existence; man may range
The court, camp, church, vessel, and the mart,
Sword, gown, gain, glory, offer in exchange
Pride, fame, ambition, to fill up his heart,
And few there are who these can not estrange;
Man has all these resources, we but one,
To love again, and be again undone.

'My breast has been all weakness, is so yet;
I struggle, but cannot collect my mind;
My blood still rushes where my spirit's set,
As roll the waves before the settled wind;
My brain is feminine, nor can forget—
To all, except your image, madly blind;
As turns the needle trembling to the pole
It ne'er can reach, so turns to you, my soul. (*DJ*, I, stanzas 193–5, lines 1537–1560, *CPW* V)

This figure of the 'trembling' needle in the compass (a spatial technology and 'instrument for showing the direction of magnetic north' [OED]) visualizes emotional teleconnections: lovers connected over distances of space and time. Like the needle turning towards the North Pole, so does Julia's soul turn to Juan. The figure itself turns on a marvellous ambiguity by which Julia's words 'my soul' are both a way of telling Juan that her soul turns to him and a form of lover's address by which Julia addresses Juan as 'my soul'.

This passage also conjures an intertextual form of teleconnection with Jane Austen's *Persuasion*, written between 8 August 1815 and 6 August 1816, its ending written and rewritten in July–August 1816, and published in December 1817 (with 1818 on the title page), just before Byron wrote this passage of *Don Juan*.[30] This passage takes us back to *Persuasion*'s eloquent form of intimate teleconnection when Anne Elliot talks to Captain Harville about women and true love, but is in fact communicating with Captain Wentworth, who is on the other side of the room, at 'the distant table', apparently out of ear-shot, busily engaged in writing a letter (to Anne, it is later revealed). Ostensibly, Captain Wentworth is writing on behalf of Captain Harville to Captain Benwick—who, as the grieving fiancé mourning the death of his fiancée by reciting copious amounts of Byron, is the text's key site of intertextual crossing with Byron. Benwick had previously entrusted Harville with the commission of a 'small miniature painting' of himself, originally 'drawn at the Cape' to give to Fanny, Harville's sister;[31] but now Benwick is asking Harville to have the miniature redirected to another woman. And Wentworth is seemingly exclusively engaged in the masculine homosocial business of re-purposing and re-directing this love token originally intended for Fanny to the other woman. A complex teleconnective circuit is set up in this scene.

First, Harville communicates obliquely with Anne, speaking to her indirectly and then directly: 'He looked at her with a smile, and a little motion of the head, which expressed, "Come to me, I have something to say;" [... then] "Look here," said he, unfolding a parcel in his hand, and displaying a small miniature painting' (*P* 252). That 'unfolding' of the parcel to display the portrait of Benwick unspools a world of teleconnections, linking Wentworth at 'the distant table'—in the same room—with Benwick's channelling of Byron, and linking the distant geographical domain of the Cape to England through the disturbances not only of war but also of heartbreak. For Harville, this is not just the death of his sister, but the death of her fiancé's love, as Benwick redirects his heart—and the love token that reciprocates the gaze of the beloved—elsewhere, bestowing that look upon another beloved. Instead of 'bringing it home for her' (*P* 252), Benwick arranges the re-direction of the portrait drawn at the Cape to another woman—Louisa Musgrove, as it turns out, the young woman whom everyone (including Anne) had thought until very recently was engaged to Captain Wentworth. It's a tangled web.

Says Harville to Anne: '"I have now the charge of getting it properly set for another! It was a commission to me! ... I am not sorry, indeed, to make it over to another. He undertakes it—(looking towards Captain Wentworth) he is writing about it now." And with a quivering lip he wound up the whole by adding, "Poor Fanny! She would not have forgotten him so soon!"' (*P* 241).

> No, replied Anne, in a low feeling voice. "That, I can easily believe". (*P* 252)

This is followed by Austen's famous illustrious-mischievous intervention about 'woman's' fickleness, literacy, education, authorship and home-boundedness.

As Anne says to Harville:

> We certainly do not forget you, so soon as you forget us. It is, perhaps, our fate rather than our merit. We cannot help ourselves. We live at home, quiet, confined, and our feelings prey upon us. You are forced on exertion. You have always a profession, pursuits, business of some sort or other, to take you back into the world immediately, and continual occupation and change soon weaken impressions. … All the privilege I claim for my own sex … is that of loving longest, when existence or when hope is gone. (*P* 252, 256)

Captain Wentworth has of course heard every word and writes his own desperate scrawl (like Donna Julia): 'I can listen no longer in silence. I must speak to you by such means as are within my reach. You pierce my soul. I am half agony, half hope.' (*P* 257–8) This maritime image of the ship's compass and Anne's piercing of Captain Wentworth's soul take us back to *Don Juan's* needle of the compass and to another exclamation of 'my soul.'

Anne exclaims—or rather Austen's free indirect narration exclaims with Anne: 'The revolution which one instant had made in Anne, was almost beyond expression. … While supposed to be writing only to Captain Benwick, he had been also addressing her!' (*P* 257). This letter to Anne, expressing Wentworth's constancy, has been written under cover of the letter of *in*constancy he had been writing for Harville to Benwick (or rather, the letter that curiously signals and ambivalently embodies both Harville's constancy as a friend and Benwick's inconstancy as a lover). This, it seems to me, is another exemplary moment of teleconnection or indirection. As John Guillory writes about indirection, in his reading of *Persuasion* and media history, 'Much communication … transmits messages already composed by another, even by the long dead.'[32] Literary texts are exemplary instances of this kind of communication through messages from the dead, over distances of space and time. This is how Byron imagined *Childe Harold* III for his daughter, Ada, who he was never to see again after her birth (when Annabella returned with her to her parents and demanded the separation), using the poem to address his child and bind her to him after his death. This is also how Mary Shelley communicated with her mother, Mary Wollstonecraft, a physically absent but intertextually structuring presence throughout Shelley's life. Shelley never met her mother, but communicated with her through her mother's writings, and her own, and the magical writings her mother read, that became the set readings for Frankenstein's creature: Goethe's *The Sorrows of Werther* (1774); Milton's *Paradise Lost* (1667); and Volney's radical Enlightenment catastrophist classic, *Ruins of Empires* (1791), which Shelley's husband adored and her father, William Godwin, despised.[33]

Byron and Austen, on the other hand, are so close, and yet so far. But here they meet at the intertextual crossing of *Persuasion* and *Don Juan*; and it is a two-way traffic. For in *Persuasion*, the conversation between Anne and Benwick about woman's constancy that *Don Juan* channels is preceded by a conversation they have about how to pronounce Byron's *The Giaour* (1813)—a conversation, incidentally, that Byron himself was still having with Murray in March 1817, while discussing *Manfred*: 'you must recollect that in the way of title—the "*Giaour*" has never been pronounced to this day' (*BLJ* 5: 192).

Austen's summer of 1816 occurs in Chawton, in the Hampshire countryside, with the drafting and completion of *Persuasion* (first completed on 18 July and then finally completed on 6 August). The drafting is stalled in the spring of 1816 under the black star of Henry Austen's bankruptcy, as well as Jane Austen's own ill-health, that had started to fail in the spring. Financial ruin and ill-health are other apocalyptic turns, then, that inform 1816 as the perfect storm of Romantic catastrophes. As Janet Todd and Antje Blank note, Austen begun the novel on 8 August 1815, the day it was generally known that Napoleon had gone into permanent exile.[34] Noting that Austen's 'adult writing career was essentially synchronous with Napoleon's military one (1793–1815)', and that a second return of Napoleon was widely feared in England, Jocelyn Harris argues for the significance of Napoleon as a structuring absence in *Persuasion*.[35] Indeed, we might read Napoleon's spectral presence into the stalled completion and double endings of *Persuasion*, as Austen ends the novel not once but twice, completing it first on 18 July 1816, and then on 6 August.

The eclipsing of Napoleon in *Persuasion* makes an interesting counterpoint to the vision of Napoleon's return staged in *Manfred*, and his dominating global presence in 1816, including an obscuring presence for Tambora. Such a shadow does Napoleon cast that he not only obscures Tambora in 1816 but had earlier appropriated the figure of the volcano for himself. As Byron had written to Thomas Moore, on 20 April 1814, about the 'Ode to Napoleon Buonaparte': "'Tis a cursed business; and, after all, I shall think higher of rhyme and reason, and very humbly of your heroic people, till—Elba becomes a volcano, and sends him out again. I can't think it all over yet' (*BLJ* 4: 100). For Byron, it is still in process—not 'over yet.' Elba did send Napoleon out again, but he went back into exile on St Helena. Written a year after Napoleon's exile to St Helena in October 1815, *Manfred* nevertheless registers this habit of Napoleonic anticipation and prediction. In Act II, scene iii, the 'Voice without, singing' predicts a second return of Napoleon from exile, aided and abetted by the Voice itself:

> The Captive Usurper,
> Hurl'd down from the throne,
> Lay buried in torpor,
> Forgotten and lone;
> I broke through his slumbers,
> I shivered his chain,
> I leagued him with numbers —
> He's Tyrant again! (*M* II.iii.16–23)

By 1817, when *Manfred* was published, this prediction had not come true, as the play dramatizes through the reference to the Bourbon restoration that came after the final defeat of Napoleon, when Nemesis refers to having been 'detain'd repairing shattered thrones, / Marrying fools, restoring dynasties (II.iii.62–63). Nevertheless, Byron himself lived in hope (just as others actively feared a return). So in this sense the Voice's prediction is more than simply an incorrect prophecy; rather, it is a self-reflexive registration of Byron's hope and/or loyalty, appropriately ironized.

Another memory text, like *Manfred*, Austen's *Persuasion* both thematizes and experiments with time: through the form of the backward romance; in the novel's concern with 'futurity' (*P* 32); in the figure of 'the ruins of the face' (*P* 78) that both delineates Anne's premature widowhood and indexes the famous fossil ruins on the shore of Lyme Regis, by the Cobb and one of 'its old wonders'(*P* 102)[36]; and in the 'general air of oblivion' by which Anne's closest friends choose to forget Wentworth. For, like *Manfred*, *Persuasion* is also a text of forgetting, of oblivion, and the sharing of a secret, which becomes the keeping of a secret:

> She was assisted, however, by that perfect indifference and apparent unconsciousness, among the only three of her own friends in the secret of the past, which seemed almost to deny any recollection of it. ... but the general air of oblivion among them was highly important ... she rejoiced anew over the conviction which had always been most grateful to her, of the past being

> known to those three only among her connexions, by whom no syllable, she believed, would ever be whispered.[37]

But that keeping of the secret, that silence, is cruel, as the narrator's irony underscores: 'perfect indifference' suggests perfect loyalty in not speaking, but also friends who don't really care. And, of course, ironically, the perfect irony and care and attention of Anne's narrator-author (unlike Anne's 'friends') is registered by the fact that she *breaks* Anne's secret by sharing it with the reader. That care and attention is enabled by Austen's formal innovations in free indirect discourse, a technique which might also be regarded as involving a particular form of space-time travel, marked as it is by a form of 'temporal shift.'[38] The free indirect style is a complex form that relies for its energies, here in *Persuasion*, on an older technology, the letter (and the older narrative first-person epistolary form), as well as on dialogue, and on its subtle shuttlings of textual quotation.

Just as Anne's most ardent interlocutor (Wentworth) is not the person she is actually speaking to (Captain Harville)—here, ironically, Anne communicates with Wentworth most intently about woman's constancy, while she is speaking face to face with *another* man—so too do Austen and Byron speak to one another not always face to face, but through their texts, and readers and interlocutors, across distances of time and space.

The intertextual conversation that Austen sustains with Byron in *Persuasion* continues in the unmarked quotation of the first line of *The Corsair* (1814), when Anne found Benwick 'drawing near her' on the Cobb, and where 'Lord Byron's "dark blue seas" could not fail of being brought forward by their present view', but where Anne's 'attention', after Louisa Musgrove's fall, 'was soon drawn per force another way' (*P* 106)—attention momentarily fixed by the present view but then drawn another way. This, it seems to me, is what happens in Austen's telegraphing of geological discussions of deep time through the reference to the 'ruins of the face' that register the ravages of Anne's odyssey of fidelity and index the fossil ruins on the shore of Lyme Regis. It also occurs in the telegraphing which is Austen's technique of quotation in the novel. These details exemplify the question of scale that is to some extent what separates but also connects Byron and Austen.

As these intertextual relations between the different worlds of Byron and Austen suggest, teleconnections between seemingly distant places sometimes lie closer to home. The matter of Austen's scale is a complex one, often underestimated by a focus on Austen's proverbial miniaturism. For, as is powerfully exemplified in the portrait miniature of Benwick, the complexities of free indirect discourse—which telescope a protagonist's interior life, but do that, as we see in the portrait and letter-writing scene, by setting that interior life in an intensely social space that is also globally connected, and by moving in and out of direct speech and dialogue, and epistolary discourse, to inflect that interior life with all the colour and movement of its broader social setting—conduct a kind of epic in miniature. Similarly, an emphasis on Austen's own closeness to home (her superb capacity for 'bringing it home') can underestimate the global circulation she achieves, not only now in the twenty-first century but earlier in the nineteenth. For just as men, sailors, pirates and Byron himself crossed 'Lord Byron's "dark blue seas"', so too did Austen's work have a reach and currency across these seas, spanning Britain, America, Canada, India, Australia and New Zealand.[39] The question of scale has always been fraught, overdetermined by Austen's own figures of the 'Two inches of Ivory' and '2 or 3 families in a country village'—figures which loom large, for better and worse, in the habitual reading of Austen's work as a form of miniaturism. But this small scale maintains an extraordinary reach, as Claudia Johnson memorably puts it in her claim that Austen's 'lean and unsparing oeuvre is like a planet on the table, not a delightful miniature, but a semiautonomous fictive world.'[40] Indeed, this semiautonomous world is created and sustained in part by the profoundly teleconnective capacities of Austen's indirect modes of intertextual quotation, epistolarity, and of the shuttling between free, indirect and direct that brings free indirect discourse home: like an epic in miniature, 'a revolution almost beyond expression', *Persuasion*'s constellating of the planets of love (and their destruction and reconstruction).

The relationship between Austen and Byron is a recurring site of magical thinking in Romantic literary history. In the context of the Tambora volcano and Alpine avalanches, it might seem to register only a slight tremor or disturbance; but it is a vital hot spot of debate and speculation in Romantic studies and has larger implications for how we think about intertextuality more broadly.

Conclusion

How do we read the moment of 1816 now? The event of Tambora, and the summer of 1816, asks us to reflect on how we think in literary studies about space and time, and how we do that in ways that sustain a conversation with science, and its subtle analysis of the space-time continuum.[41] One way of marking how 1816 continues to resonate now is to revisit the question framed by the Romantic cultural historian, Marilyn Butler, in the opening paragraph of *Romantics, Rebels, and Reactionaries* (1981), as she enquires into the rich haul of distinctive voices that marks the Romantic period: 'But why do they coincide? If they do not follow one another, what common factors caused them to develop so richly and variously at the same time? By what historical logic did Coleridge breathe the same air and read the same newspapers as Jane Austen?'[42] These questions, it seems to me, are framed under the sign of catastrophism—as is Butler's development of them: 'The great writings of the early nineteenth century are not merely pieces of historical evidence, fossils in the ground, but living texts that we too are engaged with. Just as they had no first author, they have not found their last reader. To see these works within their cultural context is also to acknowledge their place in a world we still inhabit.'[43] (And it is also to acknowledge the place that Butler's insights have in the world we still inhabit.[44]) These are the questions and reflections of a Manfred, a Cain, a Frankenstein or a Frankenstein's creature, even an Anne Elliot, that were conjured then and continue to compel now. They 'do not follow' but coincide with the impulse to rethink the space-time continuum that was inspired by the eruption of Tambora, and the teleconnections it set in motion, including the magical catastrophist writings that looked backwards and forwards from the summer of 1816.

Acknowledgement The research and writing of this essay was funded by the Australian Government through the Australian Research Council.

Notes

1. Mirka Horova, 'Byron and Catastrophism', in *Byron's Religions*, ed. Peter Cochran (Newcastle upon Tyne: Cambridge Scholars Publishing, 2011), 254. I quote from Lord Byron, *The Complete Poetical Works*, ed. Jerome J. McGann, Vol. IV (Oxford: Clarendon Press, 1986). Hereafter abbreviated *M* and referred to parenthetically in the text.

2. 'Manfred stands at the end of a path to infinite knowledge at the start of which Cain stands, thus making Byron's 1821 play what we should call a "prequel" to his 1816/17 one'; Peter Cochran, Introduction, *Manfred: An Edition of Byron's Manuscripts and a Collection of Essays* (Newcastle upon Tyne: Cambridge Scholars Publishing, 2015), 18–19. I quote here from *Byron's Letters and Journals*, ed. Leslie A. Marchand, 12 vols. (London: John Murray, 1973–1982). Hereafter abbreviated *BLJ* and cited parenthetically by volume and page number.
3. *Some Observations Upon an Article in Blackwood's Edinburgh Magazine*, in *The Complete Miscellaneous Prose*, ed. Andrew Nicholson (Oxford: Clarendon Press, 1991), 94–95. Hereafter cited as *CMP* in the text.
4. *Frankenstein, or The Modern Prometheus, The 1818 Text*, ed. Marilyn Butler (Oxford: OUP, 1998), 78.
5. Rita Felski, 'Context Stinks!' in *The Limits of Critique* (Chicago: University of Chicago Press, 2015), 162.
6. 'Teleconnections, the linkages over great distance of seemingly disconnected weather anomalies, have been identified through the appearance of geophysical processes, through statistical correlations (in space and in time) and through recognition that many atmospheric processes are manifested as waves'; Michael H. Glantz, in *Teleconnections Linking Worldwide Climate Anomalies*, ed. Michael H. Glantz, Richard W. Katz, Neville Nichols (Cambridge: CUP, 1991), 1.
7. Gillen D'Arcy Wood, *Tambora: The Eruption That Changed the World* (Princeton: Princeton University Press, 2014), 5, 8.
8. Ibid., *Tambora*, 5.
9. As Wood argues, '[t]he Tambora climate emergency followed hard upon the devastations of the Napoleonic Wars and has always remained in the shadows of that epochal conflict,' *Tambora*, 5.
10. Noah Heringman, *Romantic Rocks, Aesthetic Geology* (Ithaca: Cornell University Press, 2004), 29.
11. Ralph O'Connor, *The Earth on Show: Fossils and the Poetics of Popular Science, 1802–1856* (Chicago: University of Chicago Press, 2008), 15.
12. For an illuminating reading of Byron and catastrophism in relation to the religious politics of the larger geological debate, see Christine Kenyon Jones, '"When this world shall be former": Catastrophism as imaginative theory for the younger Romantics' in *Romanticism on the Net* (RON): 24 (Nov. 2001).
13. Wood, *Tambora*, 7–8.
14. Lyell quoted in Heringman, *Romantic Rocks*, 7.
15. Ralph O'Connor, 'Byron's Afterlife and the Emancipation of Geology' in *Liberty and Poetic Licence: New Essays on Byron*, ed. Bernard Beatty, Tony Howe, and Charles E. Robinson (Liverpool: Liverpool University Press,

2008), 147, 160. For detailed discussion of Byron as a popularizer, see O'Connor, *The Earth on Show*, 102–104, 181–182; and for quotations of Byron, see 349 (on Lyell), and 351–352 for Edward Hitchcock's quotation of *Manfred* in *Religion of Geology* (1851).

16. *Eclectic*; *Blackwood's* and *EM*, August 1817, in *The Romantics Reviewed*, ed. Donald H. Reiman (New York: Garland Publishing, 1972), ii. 961–962.
17. Horova, 'Byron and Catastrophism', 255.
18. On geology as an infidel science and challenge to Christianity, see O'Connor, *Earth on Show*, 63–69; 109–114; on geology's radical associations, see Adrian Desmond, *The Politics of Evolution: Morphology, Medicine, and Reform in Radical London* (Chicago: The University of Chicago Press, 1989), esp. 47–92. James Secord articulates the challenge of geology to Christianity with these questions: 'Who was to interpret the meaning of a science whose findings could so flagrantly be used to contradict the opening verses of the Bible? How was a scientific view to be given of the history of life which did not lead to soul-destroying materialism and atheism? Writing in these circumstances involved tactics, presenting science in a way that would calm fears of irreligion' in *Visions of Science: Books and Readers at the Dawn of the Victorian Age* (Chicago: University of Chicago Press, 2015), 138.
19. *The Gentleman's Magazine*, Jul 1817, LXXXVII, 46.
20. Jeffrey, Rev. *Manfred*, *Edinburgh Review*, Aug. 1817, XXVIII, 450.
21. *The British Review*, Jul 1817, X, 87.
22. Jeffrey, quoted in *The Works of Lord Byron*, 14 vols (London: John Murray, 1832), vol. XI, 29.
23. 'Death' is also a campy self-projection—at least it would become self-projecting after *Some Observations*, when Byron identifies 'Deaths upon pale horses' as one 'of all the things attributed to me in the last five years' of which 'I never composed nor read a syllable' (*CMP*, 89).
24. Dale Townshend neatly identifies a 'dynamic of spectral conjuration and exorcism' and a 'strategy of ghostly invitation and expulsion that is elaborated across Byron's oeuvre and which serves to frame and punctuate' the beginning and end of his career; 'Conjuration and Exorcism: Byron's Spectral Rhetoric' in *Byron's Ghosts: The Spectral, the Spiritual and the Supernatural*, ed. Gavin Hopps (Liverpool: Liverpool University Press, 2013), 98.
25. For specific details of the *Cain* controversy, see O'Connor, *Earth on Show*. See also Kenyon-Jones on how *Cain* explores the debate between William Buckland and Thomas Chalmers (who both use geology to argue for the authority of the Bible and the historicity of the Flood) and on Robert Jameson's Christian and conservative 'makeover' of Cuvier.

26. Timothy Morton, *The Ecological Thought* (Cambridge, Mass: Harvard University Press, 2010), 130–131. Such objects are associated with global warming, such as styrofoam, radioactive plutonium, and are in five kinds: viscous, molten, nonlocal, phased, interobjective.
27. *Humphry Davy on Geology: The 1805 Lectures for the General Audience*, ed. Robert Siegfried and Robert Dott (Madison: University of Wisconsin Press, 1980), 61–62.
28. On the political alignments of this divide (whereby the government and Tory periodicals featured articles by Neptunists, while the Whig *Edinburgh Review* sponsored Plutonism), see Heringman, *Romantic Rocks*, 8.
29. For a powerful argument against the conventional reading of *Childe Harold III* and *Manfred* as 'Nature poems' influenced by Wordsworth, see Jerome McGann, 'Byron and Wordsworth', *Byron and Romanticism* (Cambridge: Cambridge University Press, 2002), 173–204.
30. Jerome McGann notes that *Persuasion* is 'a book which Murray published, and which he may very well have sent to Byron, not long before this passage was written' (*CPW*, V, 680, note 1515–1516). The Sale Catalogue of 1813, item no. 154, shows that Byron owned copies of *Sense and Sensibility* (1811) and *Pride and Prejudice* (1813). See also Christine Kenyon Jones, '"He is a Rogue of Course, But a Civil One": John Murray, Jane Austen, and Lord Byron', *Persuasions* 36 (2014): 239–54.
31. Jane Austen, *Persuasion*, ed. Janet Todd and Antje Blank (Cambridge: Cambridge University Press, 2006), 252. Hereafter abbreviated P and cited parenthetically by page number.
32. John Guillory, 'Genesis of the Media Concept', *Critical Inquiry* 36 (Winter 2010): 321–362 (357, n60).
33. William St Clair, *The Godwins and the Shelleys: The Biography of a Family* (London: Faber & Faber, 1989), 340, 529.
34. Janet Todd and Antje Blank, Introduction, Jane Austen, *Persuasion* (Cambridge: Cambridge University Press, 2006), xxx.
35. Jocelyn Harris, *A Revolution Almost Beyond Expression: Jane Austen's Persuasion* (Delaware: University of Delaware Press, 2007), 73.
36. The 'wonders' of the Cobb and its view include nearby Charmouth's 'sweet retired bay, backed by dark cliffs, where fragments of low rock among the sands make it the happiest spot for watching the flow of the tide, for sitting in wearied contemplation.' Taking his title from *Persuasion*, Heringman notes that 'By embedding observations on coastal erosion and geological time in a literary description, Austen practices what I shall call "aesthetic geology"' (*Romantic Rocks*, 3).
37. *Persuasion*, 32–33.
38. Monika Fludernik argues that free indirect discourse necessitates 'temporal shift', 'when a traditional past tense context is abandoned in favour of the

present tense system, resulting in DD [direct discourse], a gnomic present tense or historical present.' 'The Linguistic Illusion of Alterity: The Free Indirect as Paradigm of Discourse Representation', *Diacritics* 25: 4 (1995): 89–115 (99).

39. On Austen as 'a producer of globally circulating texts' and 'the author of a body of texts that circulated across four continents within decades of their publication', see Melina Moe, 'Multiple Modernities in *Pride and Prejudice*' *ELH* 83. 4 (2016): 1075–1103.
40. Claudia L. Johnson, *Jane Austen's Cults and Cultures* (Chicago: The University of Chicago Press, 2012), 105. This claim is *a propos* of Johnson's parsing of Reginald Farrer 'adumbrating a theory of Jane Austen for the modernists—a lamentably understudied affiliation' (105).
41. On the inspiration that literary studies can draw from physics in its 'analysis of space-time relations', see Wai Chee Dimock, 'A Theory of Resonance', *PMLA* 112: 5 (Oct 1997), 1060–1071.
42. Marilyn Butler, *Romantics, Rebels, and Reactionaries* (Oxford: Oxford University Press, 1981), 1.
43. Ibid., 10.
44. Butler continues this exploration in *Mapping Mythologies* (completed in 1984, published posthumously in 2015), as she tracks 'so-called literary influence' in work that seems to emerges in one place, usually the metropolis, but is then seen to have 'fitted work already done elsewhere'; Butler, *Mapping Mythologies: Countercurrents in 18th-Century British Poetry and Cultural History* (Cambridge: Cambridge University Press, 2015), 93.

CHAPTER 8

Utopia or Dystopia? The Romantics in Switzerland, 1816

Steven Hampton

On Wednesday 14 August 1816, John Cam Hobhouse, close friend and former Cambridge chum of Lord Byron, approached the Swiss border near the town of Schaffhausen, after having travelled from England across France and into southern Germany. In his journal entry for that day, recorded that evening, he wrote that his party 'came to a little town in a valley, which we were told was the frontier of Switzerland which much desired land I entered without ceremony'.[1] Switzerland in 1816 was, as it had already been for centuries, a much-desired land. In the post-revolutionary, post-Napoleonic Europe of 1816, Switzerland was an idealised object of utopian desire: for many of the jaded and disappointed true believers in republicanism and democracy in Britain, the Swiss Confederation of democratic republics continued to be seen as a living manifestation of many of their political dreams.

Added to this association between Switzerland and the ideals of democracy and republicanism was, by 1816, a new appreciation of the country's physical environment. The Swiss Alps, and those of neighbouring Savoy, provided perhaps the ultimate experience of the awe-inspiring natural sublime available to European writers of the period. This natural sublimity

S. Hampton (✉)
School of Culture and Communication, Faculty of Arts, The University of Melbourne, Parkville, VIC, Australia
e-mail: steven.hampton@unimelb.edu.au

A. Collett, O. Murphy (eds.), *Romantic Climates*,
https://doi.org/10.1007/978-3-030-16241-2_8

was underwritten, in Switzerland's case, with political traditions matching the hopes and desires of radicals who had experienced the disappointment of the failure of the French Revolution and the effective repression of political dissent in Britain and Ireland. Switzerland, with its stunning mountain vistas, juxtaposed with picturesque valleys, was a place to bring together political and aesthetic longings. It was somewhere that could embody millenarian hopes, for a new heaven and a new earth, for personal redemption, for the reconciliation of humankind with the natural world and the perfecting of human society.

These were not the only associations that travellers had with Switzerland, however. The Swiss Alps had an older 'shadow side' too. For centuries, they had been feared as the abode of demons and dragons: a diabolical dystopia to be passed through as swiftly as possible on the way to the more welcoming landscapes of the Italian peninsula. This essay considers the influence which Switzerland's political and topographical environments had on the 1816-inspired works of Lord Byron and Mary Shelley, both visitors to the confederation during the 'wet and ungenial summer' of 1816.[2] In particular, it explores the duality present in some of their works: duality expressed in terms of light and dark, utopia and dystopia, mountain gloom and mountain glory.

After having spent the summer at the Villa Diodati in Cologny, Byron set off with his friend Hobhouse on a two-week trip up into the Bernese Oberland on 17 September 1816. The journal which Byron kept for his sister Augusta during this time survives, and we also have Hobhouse's diary for the corresponding period. After departing from the Villa Diodati, the men first spent two days visiting the locales on the northern side of the lake associated with Rousseau: Vevey, Clarens, Chillon and Montreux. Immediately behind these lakeside villages stands the striking, but—in comparison to the peaks of the Oberland—comparatively modest summit of the Dent de Jaman, at 1875 metres. On 19 September, Byron and Hobhouse left the lake behind, making for Montbovon in Fribourg that evening, taking in the Dent de Jaman on the way. It is an amusing episode, especially as Byron recounts it—he describes their approach to the mountain: 'steep ascent—dismounted—tumbled down & cut a finger open—the baggage also got loose and fell down a ravine, till stopped by a large tree—swore—recovered baggage—horse tired & dropping—mounted Mule'. Hobhouse went all the way to the summit but Byron didn't quite manage it. Describing the descent, Byron wrote: 'in coming down the

Guide tumbled three times—I fell a laughing & tumbled too—the descent luckily soft though steep & slippery—H. also fell—but nobody hurt'.[3]

Hobhouse's account is soberer and he tends to remark on different things than Byron. Both men, however, comment on the feeling that this episode evoked in Byron—Hobhouse writing that his friend 'observed that the glowing green pastures, with their cottages and cows in these heights, were like a dream, something too brilliant and wild for reality'.[4] In Byron's words, 'the whole of the Mountain superb—the shepherd on a very steep & high cliff playing upon his pipe—very different from Arcadia—(where I saw the pastors with a long Musquet instead of a Crook—and pistols in their Girdles)—our Swiss Shepherd's pipe was sweet—& his time agreeable'. The experience clearly made a lasting impression on both men. Byron finished his journal entry for the day as follows:

> The music of the Cows' bells (for their wealth like the Patriarchs' is cattle) in the pastures (which reach to a height far above any mountains in Britain—) and the Shepherds' shouting to us from crag to crag & playing on their reeds where the steeps appeared almost inaccessible, with the surrounding scenery—realized all that I have ever heard or imagined of a pastoral existence—much more so than Greece or Asia Minor—for there we are a little too much of the sabre & musquet order—and if there is a Crook in one hand, you are sure to see a gun in the other—but this was pure and unmixed—solitary—savage and patriarchal—the effect I cannot describe—as we went they played the 'Ranz des Vaches' and other airs by way of farewell.—I have lately repeopled my mind with Nature.[5]

Taken together, Byron and Hobhouse's descriptions of their time on the Dent de Jaman are a wonderful insight into the most common way in which Switzerland was represented and understood during the Romantic era. Hobhouse notes Byron's observation that the whole environment seemed 'too brilliant and wild for reality', like a dream, or as Byron writes at the start of that day's entry in his journal: 'the whole route beautiful as a *Dream* and now to me almost as indistinct'.[6] Byron thinks of the original Arcadia, which he had visited, and concludes that the Swiss version is preferable—it is more Arcadian than Arcadia. And he concludes that what he had seen fulfilled everything he had 'ever heard or imagined of a pastoral existence'. In *Landscape and Memory* (1995), Simon Schama observes that 'once a certain idea of landscape, a myth, a vision, establishes itself in an actual place, it has a peculiar way of muddling categories, of making metaphors more real than their referents; of becoming, in fact, part of the scenery'.[7] This is Switzerland: a mediatory space between dreams and reality, between the human and the divine, between what is and what might be.

Byron's work, of course, is not the only writing produced by the Diodati group that summer to engage with the natural and political context of Switzerland. The best-known outcome of that dreary season is Mary Shelley's *Frankenstein* (1818), partly set, like the Villa Diodati itself, on the shores of Lake Geneva, and populated by Swiss characters. It is not only that famous story which owes something to that country, however: her later novel, *The Last Man* (1826), also makes use of a vision of Switzerland as a kind of utopian alpine refuge, a setting clearly inspired by her time in that country ten years earlier. Set in the last decades of the twenty-first century, *The Last Man* details the calamitous effects of a plague which wipes out all of humanity, with the exception of four final characters, closely modelled on the Shelley ménage. In the third and final volume, the surviving main characters leave England for the continent, in the hope of escaping contagion. After arriving in a depopulated Paris, they soon decide to make Switzerland their goal, where they believe they will be safe. Having gathered together a small band of survivors, the troop eventually finds their temporary sanctuary amongst the Alps.

Switzerland's role in the story is both pivotal, because the narrative explicitly turns on it, and redemptive. In Shelley's novel, a decision is made in Paris that the survivors, now a multinational mix, will set off before the coming summer, as the plague is known to be more virulent during the warmer months, and aim to 'pass the hot months in the icy vallies of Switzerland'. Earlier in the novel, the narrator, Verney, suggests that the plague slows or disappears completely during winter, when cold, ice and snow prevent it from spreading: 'when winter came, the cold would dissipate the pestilential air'.[8] This conviction is repeated in the ensuing chapters until it culminates in the choice of Switzerland as a summer refuge. Not only is Switzerland the high-altitude home of ice and snow, it also promises both fluvial and atmospheric purity not found elsewhere: 'under the icy vaults of the glaciers, beneath the shadow of the pines … beside the streams whose intense cold proclaimed their origin to be from the slow-melting piles of congelated waters, amidst frequent storms which might purify the air, we should find health'. For many pages after the decision is made to travel to Switzerland, that land increasingly becomes an object of desire and longing for the characters, including Verney. As they prepare to set out, Verney writes that they 'looked forward to vallies lovelier than any garden, to mighty forests and halls, built not for mortal majesty, but palaces of nature's own, with the Alp of marmoreal whiteness for their walls, the sky for their roof'. It is an arduous, 'long-drawn journey' from Paris to the Swiss border, all the while the group's desperation increasing until 'to reach Switzerland, to plunge into rivers of snow, and to dwell in caves of ice, became the mad desire of all'.[9]

This 'mad desire' comes to a climax as the travellers reach the heights of the mountains of the Jura and gaze for the first time on this much-desired land:

> Nature, or nature's favourite, this lovely earth, presented her most unrivalled beauties in resplendent and sudden exhibition. Below, far, far below, even as it were in the yawning abyss of the ponderous globe, lay the placid and azure expanse of lake Leman; vine-covered hills hedged it in, and behind dark mountains in cone-like shape, or irregular cyclopean wall, served for further defence. But beyond, and high above all, as if the spirits of the air had suddenly unveiled their bright abodes, placed in scaleless altitude in the stainless sky, heaven-kissing, companions of the unattainable ether, were the glorious Alps, clothed in dazzling robes of light by the setting sun.[10]

Arrival in Switzerland promises the chance not only of physical healing but of spiritual redemption as well, via the 'spirits of the air'. The beauty of the scene so overwhelms Verney that he writes that he 'forgot the death of man'. Next to him, his companion Adrian exclaims, 'Drink in the beauty of that scene, and possess delight beyond what a fabled paradise could afford'. Another in their company cries out, 'God reveals his heaven to us; we may die blessed'.[11] Switzerland has become even more than an earthly paradise or utopia, it is heaven itself, the dwelling place of the divine: by entering it, the travellers may find salvation.

Shelley's 'Jura moment' in *The Last Man* echoes those of other writers in various European languages during the preceding decades, going back to Rousseau's novel *Julie, ou la nouvelle Héloïse* (1761) and the character Saint-Preux's exclamation upon catching a first glimpse of his homeland from the same viewpoint, after returning from years in exile. Saint-Preux experiences 'an instant of ecstasy and ravishment' as he views again 'that unique countryside, the most beautiful that ever met human eye', the sight of which restores to him 'all at once the enjoyment of my entire life'.[12] Shelley takes this a step further by adding an explicitly salvific aspect to the travellers' raptures. For Shelley's characters, nevertheless, the 'instant of ecstasy' is transitory: as soon as they cross into the land described as being 'so long the final mark and aim of our exertions', they experience disappointment and disenchantment. Verney asks, 'how could we nourish expectation of relief? Like our native England, and the vast extent of fertile France, this mountain-embowered land was desolate of its inhabitants'. Once within Switzerland, the dreams and longings of the survivors are replaced with reality and the devastating realisation that the plague has destroyed everyone here too: 'nor bleak mountain-top, nor snow-nourished rivulet; not the ice-laden Biz, nor thunder, the tamer of contagion, had preserved them'.[13]

Nevertheless, Switzerland fulfils its redemptive mission by eliminating the plague, but only after all but four of the travellers have died. The last of these other, unnamed travellers is laid to rest in 'the caves of ice beneath the glacier', the mer de glace near Chamonix, on 'a bier of ice'.[14] They are lucky to get out of the cavern alive, because an instant after they do, 'vast blocks of ice, detaching themselves from the roof, fell, and covered the human image we had deposited within'. With that inhumation in the purest of tombs, the plague is also vanquished: 'nor was it the human form alone which we had placed in this eternal sepulchre, whose obsequies we now celebrated. With this last victim Plague vanished from the earth'.[15] Switzerland enables the elimination of the plague and the salvation of humankind, but it is all too little, too late and, instead, a sense of heart-breaking loss darkens the bright ecstasies.

The summer of 2098 in the novel is a nostalgic reconfiguration of the dismal summer of 1816, which, in spite of the poor weather, was a time of unrivalled brightness and warmth in the memory of the author. *The Last Man* contains many passages heavy with the bitter-sweet memories from ten years earlier, the intervening decade having been filled with so much anguish and loss for Shelley. Between the time she spent in Switzerland in 1816 and the publishing of *The Last Man*, she lost her half-sister Fanny Imlay to suicide at the end of 1816, her one-year-old daughter Clara in 1818, her three-year-old son William in 1819, John Polidori (another of the Diodati group) to suicide in 1821, her husband in 1822 and her dear friend Byron in 1824, herself only narrowly surviving a miscarriage a few weeks before her husband's death. The novel's late summer of 2098, when the characters 'rambled through romantic Switzerland', is a brief, nostalgic sojourn, a fleeting invocation of a cherished memory in the life of the author, with a deeper undercurrent of the shared human longing for a lost earthly paradise.[16]

These kinds of utopian representations of Switzerland are, however, only one side of the coin for the Diodati group and other Romantic writers engaging with Switzerland. The dual nature of the alpine environment in the European imagination is eloquently articulated by Marjorie Hope Nicolson in her influential work, *Mountain Gloom and Mountain Glory* (1959). Nicolson explains how mountains had been viewed in both positive and negative terms throughout human history, but that, by the late seventeenth century in Europe, a more negative appraisal had become dominant, thanks especially to Thomas Burnet's work *Telluris Theoria Sacra*, or *Sacred Theory of the Earth*, published in the 1680s. Burnet expanded on the idea that mountains had not been part of the original creation and had only come into existence as a result of the deluge. This was an old interpretation of the first chapters of Genesis, which had competed with two other possibilities—that God had created mountains along with the rest of the earth's geological features on the third day of creation or that mountains had come into being immediately as a result of the fall. Burnet added the new idea of the earth being a 'mundane egg' and based his cosmology on associating the various parts of the egg with parts of the earth—Noah's flood being a result of the outer shell collapsing and the waters underneath engulfing it.

Burnet's work helped to strengthen the association between mountains and divine punishment of the earth for the sins of humankind. Nicolson quotes Thomas Heyrick, who in 1691 described mountains as the 'broken Ruines of the former World'.[17] Mountains, and for Heyrick and many other poets of his time, this especially meant the Swiss Alps, were therefore stark, prominent reminders of the fallen state of the world. In addition to the association of mountains with divine retribution came the idea that mountains somehow belonged to Satan. John Milton follows the more common Protestant line in believing that God had created mountains on the third day along with the rest of the 'dry land' and therefore does not explicitly associate mountains in *Paradise Lost* (1667) with divine punishment (in fact, there are mountains in his Garden of Eden). Despite this he nevertheless follows an old theme linking Satan to mountainous heights. In *Paradise Regained* (1671), as Nicolson points out, Milton dramatises the gospel account of the temptation of Christ by Satan atop a mountain peak. In Milton's version, Satan surveys the whole world with an 'aerie telescope' before offering all he can see to Christ. The key point here is that the mountain top belongs to Satan—it is his territory, his domain.

These sorts of associations naturally made entering mountainous areas, such as the Swiss Alps, a doubly terrifying prospect to many travellers—not only were the landscapes intimidating and filled with physical danger, they were also intimately associated with sin and evil. Many travellers' accounts of crossing the Alps, well into the eighteenth century, bear this out. The Alps were under the dominion of Satan and the dystopian dwelling place of demons, evil spirits and dragons. There are numerous instances of traditions and superstitions linking not only Satan himself but also demons, goblins, giants and dragons to various locations throughout the mountainous regions of Switzerland. Mont Blanc was actually known to the locals of Chamonix as 'Mont Maudit'—the 'Accursed Mountain'.[18] In Les Diablerets in Canton Valais, a story remains regarding devils who caused the glacier above the village to extend and envelop pastures previously used by the villagers. Other tales from the same place 'feature the mournful groaning of lost souls, whose moans heralded catastrophic landslides in 1714 and 1740'. In 1723, renowned scientist Johann Jakob Scheuchzer, professor of physics and mathematics in Zurich, published 'a compendium of all the species of dragons known to exist in the Swiss Alps, canton by canton'. Dragons were a particularly powerful symbol in the medieval Christian imagination, seen as 'winged servants and the embodiment of Satanic evil'.[19]

Wordsworth expressed a degree of exasperation at the Swiss in regard to their focus on all things diabolical. Writing in a footnote to his 1793 poem, *Descriptive Sketches*, he notes that 'the Devil with his horns, etc., seems to be in their idea, the principal agent that brings about the sublime natural revolutions that take place daily before their eyes' and that, in his opinion, 'the superstitions of the Alps appear to be far from possessing that poetical character which so eminently distinguishes those of Scotland and the other mountainous northern countries'.[20] Louis Ramond de Carbonnières, whose French translation of William Coxe's bestselling *Sketches of the Natural, Civil, and Political State of Swisserland* (1779) Wordsworth took with him on his 1790 walking tour of Switzerland, also noted this tendency amongst the Swiss in his journey over the Gotthard Pass. Not only natural features were attributed to the devil's agency but even those structures built by human hands. The road over the Gotthard included several bridges greatly admired by many travellers, one of these, the 'Devil's Bridge', described by him as possessing 'form and dimensions … equally astonishing: the span is about thirty yards wide, it rests on protuberances of two opposite rocks, and seems sustained in the air by magic'. When Ramond inquired after the name of the architect of the road and the bridges, he discovered that 'the people have forgot their benefactor, and attribute to the devil, what in their uncertainty they should have ascribed to an angel'.[21]

By the nineteenth century, at least in the wider European imagination, such diabolical ideas of the Swiss Alps had been well and truly overwhelmed by what we've already seen from Byron, Hobhouse and Mary Shelley—the alpine idyll and gateway to the sublime and the divine. But those old associations of damnation, Satanic and demonic influence still hovered around the peaks and precipices of the Swiss mountains, making the Alps an attractive setting for Gothic-Romantic works such as Byron's *Manfred* (1817), Percy Shelley's *St. Irvyne* (1811), Mary Shelley's *Frankenstein* (1818) and John Polidori's *Ernestus Berchtold* (1819).

Byron's attraction to the Satanic possibilities within the alpine landscape is clear. After climbing the Dent de Jaman on 19 September, he and Hobhouse continued into the mountainous hinterland of cantons Vaud and Fribourg, travelling on to Montbovon for the night. From there, they made their way up into the sublime high peaks of the Bernese Oberland. The environment they encountered there was of another dimension altogether from their pastoral experience at Jaman. Relatively quickly, the language each uses to describe their surroundings changes noticeably. Patrick Vincent describes Byron's alpine journal as 'written in a telegraphic style'.[22] At this point, describing their onward journey after leaving Interlaken, the central town of the Oberland, Byron's prose becomes almost breathless, even more abbreviated than before:

> entered upon a range of scenes beyond all description—or previous conception ... Arrived at the foot of the Mountain (the Yung-frau—i.e. the Maiden) Glaciers—torrents—one of these torrents nine hundred feet in height of visible descent—lodge at the Curate's—set out to see the Valley—heard an Avalanche fall—like thunder—saw Glacier—enormous—Storm came on—thunder—lightning—hail—all in perfection—and beautiful.[23]

Following this passage, Byron describes the Staubbach waterfall in Lauterbrunnen (the 900-foot torrent he's already mentioned) in an intriguing way: 'the torrent is in shape curving over the rock—like the tail of a white horse streaming in the wind—such as it might be conceived would be that of the "pale horse" on which Death is mounted in the Apocalypse'.

The next day, 23 September, Byron and Hobhouse climbed to the top of one of the peaks of the Wengernalp and saw the Jungfrau massif from one of its best vantage points. The overwhelming nature of the vistas encountered pushes Byron once again beyond language of beauty and harmony and towards something more diabolical. He writes that they 'heard the Avalanches falling every five minutes nearly—as if God was pelting the Devil down from Heaven with snow balls'. On the other side, 'the clouds rose from the opposite valley curling up perpendicular precipices—like the foam of the Ocean of Hell during a Springtide—it was white & sulphery—and immeasurably deep in appearance'.[24] Nature here is very different from the pastoral surrounds of the Dent de Jaman. Byron's experiences in the Oberland form the backdrop for his drama *Manfred* which is itself replete with Gothic imagery and diabolical tropes. This encounter with the sublime moves Byron out of the idyllic, pastoral atmosphere of the Dent de Jaman, to a place no longer comforting and at times terrifying. His temporary, comforting communion with nature is over and instead he finds himself surrounded by a natural environment which is sinister and threatening. In *Manfred*, it is this dystopian Switzerland which takes centre stage.

What *Manfred* also does is provide a kind of bridge from mountain glory—the alpine Arcadia and utopian longings—to these darker and more frightening themes of mountain gloom. At the beginning of the play, there are some fleeting hints of the Dent de Jaman, for example, when 'the shepherd's pipe in the distance is heard' and Manfred exclaims:

> Hark! The note,
> The natural music of the mountain reed—
> For here the patriarchal days are not
> A pastoral fable—pipes in the liberal air,
> Mix'd with the sweet bells of the sauntering herd;
> My soul would drink these echoes.[25]

This is the alpine utopia—and the ranz-des-vaches distracts Manfred for a brief moment, as he stands on the edge of a terrifying precipice, about to leap to his death. These pastoral thoughts quickly give way, however, and Manfred in frustration calls on the destructive power of the Alps to end his life: 'Ye avalanches, whom a breath draws down / In mountainous o'erwhelming, come and crush me'. Manfred then describes the scene before him using words little altered from Byron's alpine journal on 23 September 1816 when he and Hobhouse viewed the Jungfrau massif from the Wengernalp:

> The mists boil up around the glaciers; clouds
> Rise curling fast beneath me, white and sulphury,
> Like foam from the roused ocean of deep Hell.[26]

Manfred doesn't dive into this ocean of hell, however. Instead of his wish for annihilation being granted, 'good' Switzerland intervenes in the form of a chamois hunter, straight from the lines of Wordsworth's *Descriptive Sketches* or any other of a range of works on Switzerland from the preceding decades. The hunter stops Manfred from hurling himself into the void and takes him back to his nearby chalet.

The chalet also figures as a pivotal symbol in Mary Shelley's most celebrated work, *Frankenstein, or the Modern Prometheus*. Frankenstein's creature takes his creator to a 'hut upon the mountain', the mountain being the greatest of all the Alps, Mont Blanc.[27] Within the chalet, the creature relates his life story to Frankenstein and expresses his wish for a mate, which the doctor eventually, and reluctantly, agrees to provide. It is an exchange framed within a power relationship that resembles both the relationship between God and Satan and God and Adam in *Paradise Lost* but also distorts and subverts both of them. In *Frankenstein*, the chalet provides a space for the exchange of Satanic knowledge—knowledge which upsets the conventional order and makes a return to a previous state of understanding impossible, as Adam and Eve discovered once they listened to the serpent in the garden. For both Frankenstein and his creature, the parley in the chalet is the point on which everything pivots: afterwards, a new sense of urgency enters the narrative.

Inside the chalet in *Manfred*, a somewhat different exchange takes place: the chamois hunter attempts to reason with Manfred. Although Manfred's encounter with the chamois hunter and his time in the chalet means that he doesn't achieve his goal of immediate oblivion, he also doesn't accept the more conventional redemption offered to him within its rustic walls. Before Manfred departs, the mountaineer offers him spiritual mediation: 'my prayers shall be for thee'. But Manfred has already decided against this kind of salvation: 'I need them not … follow me not—I know my path'.[28] *Manfred*'s Switzerland is the site of an epic contest between the dark and the light, between good and evil, between the establishment and the alternative. In the end, however, neither side comes out the clear winner and both sides are rejected by the protagonist.

Manfred asserts his 'autonomy of consciousness', something which Peter Schock identifies as a key aspect of Romantic Satanism.[29] Manfred repeatedly refuses to submit to either divine or infernal authority and, by so doing, exhibits something of the independence of mind typically associated with the Satan of the Romantics, based on but not limited to Milton's Satan in *Paradise Lost.* Romantic readers of Milton were able to view Satan within that work 'as a figure misplaced and confined … in "an alien fiction"' and conclude that 'at certain points in Milton's narrative, Satan is manipulated, his behavior pulled in different ways to serve the ideological aims of the poet'. Schock gives the example of the apostrophe to the sun in Book IV, in which Satan 'considers repentance—and then abruptly rejects it, on less than credible grounds, a gesture revealing the authorial hand', as an instance of this effect.[30] As a result of this, the Romantic Satan was a figure much greater than the Miltonic original, incorporating aspects of many other characters, most notably that of Aeschylus' Prometheus. Schock documents how this Satan was used by radicals during the late eighteenth and early nineteenth centuries as a 'heroic apotheosis of human consciousness and libertarian desire'.[31] Manfred fits this mould: instead of following or obeying Satan, he is a kind of Satan himself, asserting his autonomy of consciousness.

An even more explicit example of autonomy of consciousness, achieved with a backdrop of the Satanic Alps, is portrayed by the creature in *Frankenstein.* The associations with Milton's Satan are also explicit—*Paradise Lost* is cited by the creature as one of the most influential works he has read during his conversation with Victor in the Mont Blanc chalet. He laments how inferior his situation is to that of the first created human being, Adam: 'he had come forth from the hands of God a perfect creature, happy and prosperous, guarded by the especial care of his Creator; he was allowed to converse with and acquire knowledge from beings of a superior nature, but I was wretched, helpless, and alone'.[32] Instead, the creature says that 'many times I considered Satan as the fitter emblem of my condition, for often, like him, when I viewed the bliss of my protectors, the bitter gall of envy rose within me'. The relationships between God and Satan, God and Man, Satan and Man from *Paradise Lost* are thoroughly 'satanised' in Shelley's work, becoming monstrous caricatures of the originals. Elsewhere in the novel, Frankenstein refers to the creature in satanised terms, calling him 'the filthy daemon' when he encounters him on his return to Geneva, 'daemon' being a term used 16 times throughout the novel to refer to the creature, along with 'devil' around 12 times and 'fiend' close to 40 times. Further, Justine, once she has been tried and wrongly condemned for the murder of William, expresses her horror that Elizabeth might have suspected her of being guilty of the act, calling it a crime 'which none but the devil himself could have perpetrated'.[33] So, the creature is 'daemon', 'devil', 'fiend', acts as only 'the devil himself' could and personally identifies most strongly with Milton's Satan. One of the first theatrical productions of Shelley's novel opened on 18 August 1823 at the Royal Coburg Theatre, a play by Henry M. Milner titled *Frankenstein, or the Demon of Switzerland*.[34] Switzerland and the Alps were a particularly fitting home for Frankenstein's creature and he is able very effectively to embody, to distil some of the diabolical presence inhabiting the Swiss mountains. At the same time, his creator, Victor Frankenstein, also exhibits a Satanic autonomy of will in his very act of creation.

As for *Manfred*, in Schock's view, the culmination of the drama is 'the hero's Miltonic speech to his infernal "genius"', in which he 'declares the autonomy of "the mind which is immortal"—that is, the mind of Manfred alone'.[35] Furthermore, Schock demonstrates how the Romantics used Satan to 'mythicize the human struggle against various forms of oppression and limitation', thereby gaining a 'mythic medium for articulating the hopes and fears their age aroused, for prophesying and inducing change'. Before them, in the 1790s, William Godwin, amongst others, identified Milton's Satan as 'an embodiment of the fully autonomous intellect that discerns and rejects the radical injustice of a "despotic" and "assumed" power analogous to the arbitrary authority of prescription and precedent that governed England'.[36]

This understanding of Satan as an autonomous force, as opposed to the established power of a 'despotic' divine ruler, finds indirect expression in *Frankenstein*, through the interchange between light and darkness which occurs throughout the novel and acts to frame the moral and political questions raised by the story. More than once, Shelley creates such an effect using the alpine landscape—a contrast is made between the Jura Mountains on Switzerland's western border with France and Mont Blanc to the southeast, at the other end of Lake Geneva. As Frankenstein returns to Geneva from Ingolstadt after receiving news that his youngest brother, William, has been murdered, he recounts how he views the mountains as he approaches his home and 'discovered more distinctly the black sides of Jura, and the bright summit of Mont Blanc'.[37] At the sight, he 'wept like a child' and cried out 'Dear mountains! my own beautiful lake! how do you welcome your wanderer? Your summits are clear; the sky and lake are blue and placid. Is this to prognosticate peace, or to mock at my unhappiness?'[38] Good and evil, comfort and mockery, are simultaneously present in the topography of the region. Later, just after Victor marries Elizabeth and relates that 'those were the last moments of my life during which I enjoyed the feeling of happiness', he goes on to describe the scene from the shores of the lake, 'the beautiful Mont Blanc, and the assemblage of mountains that in vain endeavour to emulate her', and on the opposite side, 'the mighty Jura opposing its dark side to the ambition that would quit its native country, and an almost insurmountable barrier to the invader who should wish to enslave it'. Mont Blanc is a peak towards which our attention and our longing are drawn, heights towards which our ambition aspires. The Jura is the opposite—it limits and contains and prevents exchange. This representation of these opposing mountain massifs, at either end of the *pays de Vaud* and the lake, echoes God and Satan, good and evil, glory and gloom, but, curiously, reverses our expectations. It is God, the God of the establishment, who limits and contains human ambition, just as the black Jura does to those trying to leave or enter Switzerland. Mont Blanc, on the other hand, is the revolutionary angel of light, Lucifer, calling to his human subjects to follow him and emulate his autonomy. The presence of these mountains amplifies the sense of the country between them—the *pays de Vaud*, Saint-Preux's 'unique countryside, the most beautiful that ever met human eye'—being contested land, an earthly paradise, a Garden of Eden over which two titans struggle.

Shelley connects this cosmic contest, with its strong echoes of *Paradise Lost*, back to the more earthbound contest continuing, albeit in a more muted fashion, in the revived post-Napoleonic monarchies of Europe when her novel was published—a contest between republican and monarchical ideologies begun in earnest in England in Milton's time. There are several references in *Frankenstein* to Switzerland's republican heritage and institutions, such as in the letter from Elizabeth to Victor explicitly extolling their native country's republican and democratic advantages: 'the republican institutions of our country have produced simpler and happier manners than those which prevail in the great monarchies that surround it' and 'a servant in Geneva does not mean the same thing as a servant in France and England'.[39] Aligning Switzerland's natural environment and physical features with its political nature and practices creates a mutually amplifying loop: increasing the sense of imposing sublimity in the landscape, while simultaneously lending a material gravity to Swiss republicanism.

This topographic-political consolidation can be articulated using a Satanic paradigm just as validly as it is more frequently used by Romantic writers to associate divine approval with landscape and political goals. Not only can Switzerland support a Satanic political agenda, it can be seen as a kind of topological embodiment of Satan himself. The many explicit names in the Alps attributing various geological features to the devil contribute to this, but the parallel is also valid in a political sense. Romantic Satanism frequently used the person of Satan to express autonomy of will or consciousness, independence of character and implacable opposition to established, hierarchical power structures. Switzerland is used during the Romantic period to express all of these things too. The Swiss alpine peasants of Rousseau, Wordsworth and others are the epitome of the autonomy of will which contemporary authors also expressed through descriptions of Satan, a figure, as already noted, only loosely based on the Miltonic original. Further, these same peasants, and the Swiss more generally, were seen to be of entirely independent character—possessing social mores and institutions which were unique within Europe, such as their oft-cited sumptuary laws and their celebrated democratic community gatherings, the *Landsgemeinden*. And no nation or society in Europe was as clearly depicted as being opposed to forms of social or political hierarchy as Switzerland: examples of this abound in the works of virtually all authors writing about Switzerland, regardless of genre, whether in the travel journals of Coxe or Ramond, the poetry of Wordsworth or the at-times-comical portrayal of egalitarian Swiss characters in Walter Scott's novel, *Anne of Geierstein* (1829). Throughout the Romantic era, perhaps especially in the post-Waterloo world of 1816, there was no place in the European imagination which so consistently embodied political autonomy and rejection of despotic or assumed power as Switzerland. Switzerland could not only be used to support a Satanic political agenda, it provided a geo-national and socio-political embodiment of the same key attributes which contemporary, predominantly radical, writers also projected onto Satan.

Switzerland, then, is as much the home of a diabolical dystopia as it is the seat of a divine utopia. As Wordsworth observed, the inhabitants of the Swiss Alps saw Satan as being 'the principal agent' at work in their homeland. Throughout the centuries leading up to the Romantic period, the Swiss Alps were seen as a place of terror and Satanic hegemony, and it was really only during the eighteenth century that the alternative vision of a sublime, pure, divinely blessed alpine environment became dominant.

Although Romantic writers are more usually associated with this later image of Switzerland, there are still numerous Romantic works which make use of a dark, terrifying depiction of the alpine land. Simultaneously, Switzerland is also used in support of a radical, heroic Satanism. In all of this, there is a clear duality: Switzerland, the much-desired land, can be deployed equally well on the side of good or evil, light or dark, establishment or alternative, status quo or change.

Notes

1. John Cam Hobhouse, *Diaries of John Cam Hobhouse* (British Library Add. MS.56536, 1816), 67.
2. Mary Wollstonecraft Shelley, Ronald Levao, and Susan J. Wolfson, *The Annotated Frankenstein* (Cambridge, MA: The Belknap Press of Harvard University Press, 2012), 332.
3. George Gordon Byron and Leslie A. Marchand, *Byron's Letters and Journals*, ed. Leslie A. Marchand, 12 vols., Vol. 5 (London: J. Murray, 1973–1982), 98, 99.
4. John Clubbe, Ernest Giddey, and John Cam Hobhouse, *Byron et la Suisse: Deux Études*, Publications de la faculté des lettres, Université de Lausanne (Genève: Droz, 1982), 42.
5. Byron and Marchand, 99.
6. Ibid., 98.
7. Simon Schama, *Landscape and Memory* (New York: A.A. Knopf: Distributed by Random House, 1995), 61.
8. Mary Wollstonecraft Shelley, Jane Blumberg, and Nora Crook, *The Last Man*, vol. 4, *Novels and Selected Works of Mary Shelley* (London; Brookfield, VT: W. Pickering, 1996), 298, 155.
9. Ibid., 306, 310, 311.
10. Ibid., 324.
11. Ibid., 324, 325.
12. Jean-Jacques Rousseau, Philip Stewart, and Jean Vaché, *Julie, or, the New Heloise: Letters of Two Lovers Who Live in a Small Town at the Foot of the Alps*, ed. Roger D. Masters and Christopher Kelly, 13 vols., vol. 6, *The Collected Writings of Rousseau* (Hanover: Dartmouth College: University Press of New England, 1997), 344.
13. Shelley, Blumberg, and Crook, 4, 327.
14. Technically, Chamonix and Mont Blanc are just outside of Switzerland, but they were often treated as part of it by travellers during this period. Shelley, Blumberg, and Crook, 4, 329.
15. Ibid., 329, 330.

16. Ibid., 331.
17. Heyrick quoted in Marjorie Hope Nicolson, *Mountain Gloom and Mountain Glory: The Development of the Aesthetics of the Infinite*, Weyerhaeuser Environmental Classics (Seattle: University of Washington Press, 1997), 230.
18. Andrew Beattie, *The Alps: A Cultural History*, Landscapes of the Imagination (New York: Oxford University Press, 2006), 109.
19. Ibid., 110, 113, 115.
20. William Wordsworth and Eric Birdsall, *Descriptive Sketches*, ed. Eric Birdsall and Paul M. Zall, 21 vols., The Cornell Wordsworth (Ithaca, NY: Cornell University Press, 1984), 84.
21. French: le Pont du Diable, German: die Teufelsbrücke. William Coxe and de Carbonnières Louis François Élisabeth Ramond, *Travels in Switzerland, and in the Country of the Grisons*, 3rd ed., 3 vols., vol. 1 (Basil: James Decker, 1802), 280.
22. That is, 'écrit dans un style télégraphique'. Patrick Vincent, *La Suisse vue par les écrivains de langue anglaise*, Le savoir suisse (Lausanne: Presses polytechniques et universitaires romandes, 2009), 96.
23. Byron and Marchand, 100–101.
24. Ibid., 101, 102.
25. George Gordon Lord Byron, *The Complete Poetical Works*, ed. Jerome J. McGann, 7 vols., vol. 4, Oxford English Texts (Oxford; New York: Clarendon Press; Oxford University Press, 1980), ll.46–52.
26. Ibid., ll.75–76, 85–87.
27. Shelley, Levao, and Wolfson, 174.
28. Byron, *Complete Poetical Works*, ll.89, 90–94.
29. Peter A. Schock, *Romantic Satanism: Myth and the Historical Moment in Blake, Shelley and Byron* (Houndmills, Basingstoke, Hampshire; New York: Palgrave Macmillan, 2003), 37.
30. Ibid., 35.
31. Ibid., 39.
32. Shelley, Levao, and Wolfson, 209.
33. Ibid., 210, 139, 150.
34. See ibid., 372.
35. Schock, 97.
36. Ibid., 36, 6, 2.
37. Shelley, Levao, and Wolfson, 137.
38. Ibid.
39. Ibid., 284–285, 126.

CHAPTER 9

Metaphor and the Unprecedented: Byron's 'Darkness' and Responding to Ecological Disaster

James Phillips

George Gordon Byron's 'Darkness' (1816) is a poem about the end of the world. For that reason, it cannot make sense: it refers to its own impossibility of reference. The poem has to buck against our tendency to find a place for the unprecedented within the continuity, such as it is, of our meaning-making practices. 'Darkness' applies itself to the literal description of the extinction of the sun, global slaughter and cannibalism, the withering of the winds and the laying waste of forests. Even though its focus remains steadfastly on this universal cataclysm, the poem has again and again been read metaphorically as though it cannot mean what on first impression it appears to mean. A manifold strategy of recuperation is in play.

Walter Scott, in an early review, deplores what he obtusely considers the poem's lack of reference: 'In this case our author has abandoned the art, so peculiarly his own, of shewing the reader where his purpose tends, and has contented himself with presenting a mass of powerful ideas unarranged, and the meaning of which we certainly confess ourselves not

J. Phillips (✉)
School of Humanities and Languages, University of New South Wales,
Sydney, NSW, Australia
e-mail: j.phillips@unsw.edu.au

A. Collett, O. Murphy (eds.), *Romantic Climates*,
https://doi.org/10.1007/978-3-030-16241-2_9

always able to attain.'[1] Unwilling to make nothing whatsoever of the poem and contending that its failure of reference nonetheless illuminates the dangers to which an unbridled imagination exposes a poet, Scott paternalistically recommends to its author the comforts of resignation: the alleged misfire of the poem becomes a metaphor for the arrogance of the romantic outsider.[2]

Gillen D'Arcy Wood is one of several recent commentators who, sharing Scott's dissatisfaction with a 'naïve' interpretation of the poem, confidently claim to discover its referent in the aftermath of the 1815 eruption of Mount Tambora. He writes: 'Byron's "Darkness" has routinely been misread as an apocalyptic allegory. Actually, it is a literary speculation on a literal event, a profoundly ecological poem in its intuition of both the human impact of natural disaster … and its harrowing images of an environmentally degraded world.'[3] Neither Scott's nor Wood's exegesis is unmotivated. In support of the claims he makes with regard to Byron's character, Scott adduces in addition to 'Darkness' other texts its author had already published by the time of the review. For his part, Wood mentions the composition history of the poem: recalling the summer he spent at the Villa Diodati near Geneva, Byron himself identified the inspiration of 'Darkness' as the meteorological disturbance, since attributed to the Tambora eruption, 'when there was a celebrated dark day, on which the fowls went to roost at noon, and the candles were lighted as at midnight'.[4] These two interpretations of the poem, as much as their respective subjectivism and objectivism appear to establish them as mutually exclusive, are alike in their unease with the surface meaning of 'Darkness'. While an unease with the idea of the self-sufficiency of the literal is, from one vantage ground, the sustaining affect of all commentary, a question arises in relation to Byron's poem that concerns the ethics and probity of this unease. Whenever language's stock of existing names comes up short in the confrontation with the new, metaphor is a means available for folding the unfamiliar into the familiar by stretching the semantic coverage of terms. As Max Black writes in his influential intervention in the philosophy of language: 'Metaphor plugs the gaps in the literal vocabulary (or, at least, supplies the want of convenient abbreviations). So viewed, metaphor is a species of *catachresis*, which I shall define as the use of a word in some new sense in order to remedy a gap in the vocabulary; catachresis is the putting of new senses into old words.'[5] Yet this pragmatics of metaphor hits a limit in the case of the absolutely unprecedented—and, by definition, the end of the world has not happened before and therefore resembles nothing. Here metaphor ceases to be a tool and becomes an opiate.

Of course, Byron has no right to defy his interpreters in this way, for he has not literally described the end of the world. As his testimony is, strictly speaking, incredible, there is in his words that 'semantic impertinence' which Jean Cohen was to analyse as an invitation to metaphorical readings.[6] Semantic impertinence, whereby an utterance resists accommodating conventional usage, primes a reader or listener to admit novel applications of terms: what is Byron's text about if it cannot be about the future end of the world? 'Darkness' pointedly does not announce itself as a mere premonition of the end of the world. Its use of the past tense throughout is consistent with the pretension to describe, rather than simply prophesy, just as the lines 'And War, which for a moment was no more, / Did glut himself again' appear to situate the events described in the brief period since Napoleon's defeat at Waterloo. Not only does 'Darkness' profess to relate a series of incidents on so catastrophic a scale that they could not all escape the attention of Byron's contemporaries, it also delineates conditions in which the narrator's biological survival and hence the very possibility of composing the text is inconceivable. If Byron's poem begins by gesturing towards fantasy, the gesture is nonetheless immediately retracted: 'I had a dream, which was not all a dream.' While 'Darkness' eschews the relative innocuousness of dreams, it likewise acknowledges that it cannot pass muster as a dispassionate chronicle of recent events. This undecidability is at the heart of the poem's semantic impertinence. Its words grope airily for a referent like the tendrils of a creeper. As 'Darkness' does not name its reality directly, it prompts commentators to search out just how much reality there is adhering to it: for Walter Scott, the tangible substrate of the poem is the turbulent imagination of the poet, whereas for Gillen D'Arcy Wood it is the planetary impact of the eruption of an Indonesian volcano. Only once it is bolstered by these supplementary referents can 'Darkness' take on the hue and functionality of a metaphor. The difference between interpretations suggests that authorship of the incompatible and recondite metaphors purportedly constituting the poem does not rest solely with Byron, but involves an ex post facto collaboration with his critics.

If the reference of a text is deemed to be so unclear as to justify the specialist labour of a commentator to draw it out, it can be asked whether the exertion on show in the critical enterprise is not due to the kicking and screaming with which a text struggles against being made to deliver up a metaphor. This is not to imply that Byron knew or was even able to know that the Tambora eruption was the cause of the climatological anomalies of the summer of 1816 and deliberately disavowed this knowledge in the writing of his poem (while Benjamin Franklin speculated as early as 1784 whether worldwide atmospheric changes were the result of volcanic events, the link was first seriously argued only in 1888 by the Krakatoa Committee of the Royal Society).[7] Needless to say, noting a text's taciturnity towards the metaphors it is alleged to harbour is not the same as valorising the surface meaning and the author's intention dubiously equated with it. A text cannot escape the production of metaphors to which the act of reading exposes it, but insofar as there is a distinct friction between 'Darkness' and the senses ascribed to it, it is worth querying what this text's overt aporia of meaning-making says about both the vocation of poetry and the business of interpretation in the face of environmental catastrophe.

The choice of the simple past tense is central to the hermeneutic perplexity and outrage of 'Darkness'. By presuming to occupy a position from which not only to observe the end of the world but also to recollect it, the narrator presses the case for the aloofness and independence of the romantic poet to the point of absurdity and self-contradiction (in 1816 at his most receptive to what he regarded as Percy Bysshe Shelley's 'metaphysical' vein, Byron seeks here to outdo him in abstraction). The act of description is not assigned to an anonymous agency. Beginning with the first-person singular pronoun, the poem does not elaborate the identity of its narrator beyond the bare ability to narrate the extinction of the universe. After the initial line the subject of every verb is in the third person. By declining to alternate between description and confessional statements regarding his response to what he sees, the narrator leaves it up to the apocalypse's thematic materials to move a reader. This authorial restraint, apart from flagging an awareness of the bathos that lies in wait for literary invocations of the end of the world, intimates that with the dimming of the sun and stars the poet's powers of emotional response are also dimmed (a generation later, in Georg Büchner's *Woyzeck* the rhetorical register of the apocalypse is scaled down still further to that of a children's bedtime story).[8]

Judged, however, as a poem on the Year Without a Summer (and Byron's own account of the poem's genesis encourages such a judgement without authorising it), 'Darkness' is anything but restrained. The crop failures, famines, flooding and cholera outbreaks that are among the recorded effects of the Tambora eruption pale—one might say:

reprehensibly—into insignificance beside the poem's vision of the darkening of the universe. The weather event that Byron witnessed at the Villa Diodati becomes the pretext for the mobilising of an array of hyperbolic literary tropes. An early commentator on the poem, writing in 1828 in a letter to *The Imperial Magazine*, proposes a series of verses compiled from Jeremiah, Isaiah and Ezekiel as Byron's Biblical models.[9] By resorting to imagery familiar from the Old Testament so as to craft a response to ecological disaster, Byron inserts the exceptional circumstances of 1816 into the longstanding frame of Judeo-Christian discourse: the sulphuric acid aerosol clouds of the Year Without a Summer are thus conscripted for service as a novel metaphor for the apocalypse—a way of giving a new lease on life to an old trope. From the perspective of environmental observation, there is accordingly a levity, even flippancy, in the overblown nature of the incidents that Byron chooses to describe. The aftermath of the Tambora eruption is not seen for what it is because Byron allows literary precedents to block his view and to upstage reality.[10]

Morton D. Paley has stressed that the parallels available to be drawn between 'Darkness' and various Biblical verses do not cancel out Byron's divergences from religious tradition:

> The invocation of apocalypse leads to the expectation of a millennium. 'Darkness' does indeed raise for a moment the vision of a world in which, in the terms of Isaiah, 'they shall beat their swords into plowshares, and their spears into pruning-hooks: nation shall not lift up sword against nation, neither shall they learn war any more' (2: 4). However, this hope is brutally dissipated. ... Millennial associations are consistently invoked in order to be bitterly frustrated, as they would also be in Campbell's 'The Last Man'. In the year without a summer, which was also the year after the Congress of Vienna and the reinstitution of the old political order in Europe, nature must have seemed to parallel the world of human events. Unlike apocalyptic poems of the 1790s like Coleridge's 'Religious Musings' and Blake's *America*, 'Darkness', a true product of post-revolutionary awareness, presents apocalypse but no millennium.[11]

To charge the poem with flippancy in its response to ecological disaster is by no means to call into question the seriousness or even the necessity of its aggression towards Christian millennialism. With its opening line 'Darkness' stakes a claim to a foothold in existing conditions. Against the spate of restorations ushered in by the fall of Napoleon, it ranges its poetic vision of a universe whose every light has sputtered out. From the weather event caused by the Tambora eruption, it distils not a warning of what might happen but rather a chronicle of what has already come to pass. Having become legible in the sky itself, the death of political hope ineluctably encompasses the demise of religious hope.

Interpreted as an instance of how European literature attempted to come to terms with, to contest and to elucidate the change in political fortunes after 1815, 'Darkness' has recourse to Biblical tropes in order to draw attention to an experience of the present that Byron's contemporaries might otherwise have denied sharing. According to this reading of the poem, Byron is engaged in the construction of a metaphor. He has to do violence to the Biblical tropes before they will give up a new sense and name a phenomenon—the irremediable hopelessness of life here and now—for which language has no words already in circulation. To dislodge the Biblical tropes from their long-established circuits of meaning, Byron hitches them to the meteorological disturbance of the Year Without a Summer, which in turn survives in the poem only by being thoroughly misrepresented. If the poetic undertaking of 'Darkness' conforms to the conventions and goals of neither natural science nor scripture scholarship, it is because the text plays fast and loose with both disciplines in scrabbling for the means to name something that eludes the semantic extension of all previously coined terms. The metaphor of 'Darkness' is the resemblance it is said to uncover between the climatic conditions of 1816, the political mood of the Europe of the day and the Judeo-Christian apocalypse.[12] The metaphor has traction only so long as this resemblance is conceded. Yet it also retains the status of a metaphor only so long as the resemblance is not taken for granted: were the lines of communication between the aftermath of the Tambora eruption, the politics of 1816 and the end of the world to become a matter of general agreement, then the juxtaposition of the three would cease to be a distinctly poetic act.[13]

As though to shore up the appearance of poetic agency, the first verse of 'Darkness'—'I had a dream, which was not all a dream'—affirms the presence of the work's author and the personal quality of his vision. What this 'I' proceeds to recount is not to be mistaken for actuality (all efforts to catch the poem out in its falsifications are therefore embarrassingly maladroit), even as it refuses to be dismissed as the irrelevant fantasia of a lone individual. The poem is a description of the world that discloses itself within a particular mood, namely, the mood of despair. How well it describes this world cannot be evaluated by criteria associated with other moods. For other moods (and the equanimity that underpins the dispassionate observation of phenomena is itself a mood), the world did not come to an end in 1816. The truth of despair is its perception that the universe has always already extinguished itself. While this truth is not communicable to other moods, it does not follow that it is not communicable at all. Byron's poem enters into its intelligibility within the communion of the despondent: it meets with intersubjective confirmation from those readers to whom the mood of despair discloses its truth. However unserviceable this monolithic and unresponsive truth no doubt is for navigating a world that has not come to an end, the detailed knowledge of what is when dealing, inversely, with a world that has ceased to exist is comparably useless.

The Year Without a Summer is not, for Byron, the first sign of the physical untenability of the known universe nor does his despair have its origin in the waning prospects for political revolution in post-Napoleonic Europe. In the timeliness that was immediately imputed to Byron's poetry there is an opportunism, since this reception of his writings involves a translation of his private afflictions into public concerns. Yet as Byron's despair is inseparable from his literary admirations, the ostensibly personal points beyond itself. In two poems from Byron's juvenilia, 'Ossian's Address to the Sun' (1805) and 'A Version of Ossian's Address to the Sun' (1806), the failure of the sun in 'Darkness' is presaged, but whereas in the later poem it is tied to the observation of a meteorological phenomenon in the skies over Lake Geneva, here it is more a matter of extrapolation from James Macpherson's crepuscular thematics. The Year Without a Summer, when it comes to pass, finds a niche already prepared for it in Byron's poetry: it is fit for purpose for the manifestation of certain states of mind and for the deployment of certain literary effects. To put it in this way is to suggest that there is an asymmetry between the vehemence of Byron's response to the weather event that he witnessed in the summer of 1816 and his actual registration of its unprecedented character (unprecedented, i.e. in the living memory of Byron and his contemporaries). 'Darkness' is the moment not so much when English poetry opens itself up to the gravest emergencies of ecocriticism as when for its own ends it pretends to do so. But precisely by attending to its own affairs, 'Darkness' nonetheless broaches questions concerning the relationship of human agency to environmental catastrophe.

At issue is the condition of possibility of a spectatorship of the end of the world. The bare 'I' of 'Darkness', which by unexplained means manages to stand apart from the general cataclysm, thereby observing it relatively unharmed, is the alienated consciousness that disavows its own material basis. With its vision of an apocalypse in which the abstract subject is not implicated, 'Darkness' pays the highest conceivable compliment to the independence of human thought from its environment. But the flattery of the poem is so disproportionate and preposterous that its recipient is clearly the target of mockery and critique. The assertion of human autonomy in the early modern period has become in Byron's romanticism the mourning over an unbridgeable gulf between the self and the world. The feeling of invulnerability in the face of the power of nature, which Immanuel Kant treats in his 'Analytic of the Sublime' (1790) as a source of pleasure, is now a source of pain. The annihilation of the world in 'Darkness' is one way in which Byron gives poetic form to this unbridgeability, just as the title character's expansive speeches on guilt and cosmic isolation in *Manfred* (1817) are another. Timothy Morton, relating Byron to the exposition of the Unhappy Consciousness in G. W. F. Hegel's *Phenomenology of Spirit* (1807), discerns in this mourning over an unbridgeable gulf a rejection of the complacency and putative independence of consciousness. A poem such as 'Darkness' is, for Morton, accordingly marked less by a failure to do justice to the volcanic winter that was its stated inspiration than by its revelation of the insolvency of the abstract subject: 'We can recognize this soil as the very poisoned ground on which we are actually living at this moment. It is the ground that modernity, emerging from the age of reason, is successfully turning into something like the effects of a nuclear catastrophe, without even having to use nuclear bombs.'[14] The 'I' of 'Darkness' is able to perceive the end of the world in the Year Without a Summer because in the agony of its narcissism it only ever meets with reflections of itself.

Byron's poem does not *predict* that modern subjectivity in its indifference to its ecological conditions of possibility will blithely bring about the end of the world (prediction implies a future occurrence that, as it has yet to befall the modern subject, may lie in the latter's power to prevent). The disaster has already taken place. The 'I' that opens the poem has no sphere of activity apart from the contemplation of the devastated world that is its essential correlate. For Jerome McGann, the poem whittles away the abstract subject's appearance of agency: '"Darkness" records a process of (as it were) *un*discovering the powers of the human mind.'[15] That which McGann aligns with impotence is not irreconcilable with the responsibility for global catastrophe that Morton lays at the door of modern consciousness. And that the two are not mutually exclusive, but on the contrary interdependent, is the well-known position that Martin Heidegger was to maintain from the 1930s onwards in his polemic against the metaphysics of subjectivity of technological civilisation. In his interview with *Der Spiegel*, Heidegger speaks of the powerlessness of human beings with respect to the tyranny of their own agency: only another god or the absence of god can henceforth save us.[16]

Byron's 'Darkness', in many respects, reads like a precursor of the apocalyptic strain in Heidegger's own writings of more than a century later. Heidegger's critique of the worldlessness of the modern subject has more purchase on the Byronic persona than on the philosophical tradition that he takes as his nominal interlocutor (Byron's Manfred is arguably closer to the caricature of Descartes at which *Being and Time* aims its barbs than Descartes himself). The 'I' of 'Darkness' is too disengaged from the world to compose a poem in which the volcanic winter that resulted from the Tambora eruption retains its features and scale. Notwithstanding its gravity, the event becomes an occasion for Gothic metaphorising and poetic ventriloquism. Natural phenomena are not trusted to speak for themselves: the 'I' surveys a desolate world from which it is independent but which, conversely, is not independent of it. The imaginative flights of 'Darkness' depict a catastrophe that has been humanised, as it were. If 'Darkness' is deficient as a poem about the Year Without a Summer, it is because its horizon of signification is determinedly anthropogenic. It is able to appear to foretell the disaster of twenty-first-century global warming because it is not able to break out of the bewitched circle of aggrandised romantic subjectivity: that this is a source of consternation for Byron is more to his credit than any gift of clairvoyance that might be notched up to him.[17] 'Darkness' is a poem about the end of the world that is the narrative translation of the worldlessness of the abstract subject. It is the metaphor that condemns the facility with which the romantic self, in its independence of the givenness of what is, is able to manipulate phenomena into metaphors. The volcanic winter that followed the Tambora eruption, in being made to say something other than itself, retreats from view to leave the 'I' with nothing but the lightless void.[18]

'Darkness' is a metaphor for the limits on the semantic elasticity of terms. It stages the failure of language to accommodate the unprecedented. Gripped by despair and its foreclosure of the future, the world it envisions is 'A lump of death—a chaos of hard clay'. If the end of the world is the despair of language and the impossibility of sense, a poem that addresses it will also be unable to address it: it will necessarily be understood to refer to something else. To see in the volcanic clouds of 1816 a resemblance to the end of the world is not so much Byron's talent as the problematic cast that he gives to the freedom of the imagination and its meaning-making.

Notes

1. Walter Scott, '1. *Childe Harold's Pilgrimage*, Canto III. 8vo. 2. *The Prisoner of Chillon, a Dream; and Other Poems.* By Lord Byron. 8vo. John Murray: London', *The Quarterly Review* 16.31 (1816): 172–208, 204.
2. Ibid., 207–209.
3. Gillen D'Arcy Wood, 'The Volcano Lover: Climate, Colonialism, and the Slave Trade in Raffles's "History of Java" (1817)', *Journal for Early Modern Cultural Studies* 8.2 (2008): 33–55, 44.
4. George Gordon Byron, *His Very Self and Voice: Collected Conversations of Lord Byron*, ed. Ernest Lovell, Jr. (New York: Macmillan, 1954), 299.
5. Max Black, 'Metaphor' in id., *Models and Metaphors: Studies in Language and Philosophy* (Ithaca, NY: Cornell University Press, 1962), 25–47, 32–33. Compare this with the praise of metaphor in Aristotle, *Poetics*, trans. W. H. Fyfe (London: Heinemann, 1927), 1459a: 'By far the greatest thing is the use of metaphor. That alone cannot be learnt; it is the token of genius. For the right use of metaphor means an eye for resemblances.'
6. Jean Cohen, *Structure du langage poètique* (Paris: Flammarion, 1966).
7. F. A. Rollo Russell and E. Douglas Archibald, 'Part IV. On the Unusual Optical Phenomena of the Atmosphere, 1883–1886, including Twilight Effects, Coronal Appearances, Sky Haze, Coloured Suns, Moons, &c.' in Ralph Abercromby et al., *The Eruption of Krakatoa and Subsequent Phenomena: Report of the Krakatoa Committee of the Royal Society*, ed. G. J. Symons (London: Trübner, 1888), 151–463. While the global scale of volcanic activity was not scientifically defended before the late nineteenth century, the eclipse-like conditions that pyrocumulus clouds could bring about in the vicinity of eruptions were longstanding common knowledge. See the description of his uncle's death during the 79 CE eruption of Vesuvius in Pliny the Younger, *Complete Letters*, trans. P. G. Walsh (Oxford: Oxford University Press, 2006), 145: 'They used strips of cloth to fasten pillows on their heads as a protection against falling stones. By now it was daylight elsewhere, but there it was night, blacker and denser than any night.'
8. See Georg Büchner, *Woyzeck* (1879) in id., *Danton's Death; Leonce and Lena; Woyzeck*, trans. Victor Price (Oxford: Oxford University Press, 1971), 105–132, 128: 'Once upon a time there was a poor little boy who

had no father or mother. Everything was dead, and there was nobody left in the whole wide world. Everything was dead, and he went away and searched day and night. And because there was nobody left on earth he thought he'd go up to heaven. And the moon looked at him so kindly! But when he reached the moon he found it was a piece of rotten wood. And then he went to the sun, and when he reached the sun he found it was a withered sunflower. And when he came to the stars they were little golden gnats that a shrike had stuck on a blackthorn. And when he wanted to go back to earth, the earth was an upturned pot. And he was all alone. And he sat down and cried, and he's sitting there still, all alone.' The disillusionment characteristic of the *Vormärz*, which in Europe coloured the early reception of Byron's lyric poems as timely articulations of the general mood, had by the mid-1830s seeped even into children's literature, as though no sphere of human activity, according to Büchner, could be considered spared the consequences of the rout of revolutionary ideals.

9. R. L. L., 'The Bible and Lord Byron', *The Imperial Magazine* 10.116 (1828): 699–701.
10. This literary mediation of Byron's descriptions of nature has long been a topic of dispute. See Wordsworth's comment on *Childe Harold*, III as recorded in the entry for 27 October 1820 in *The Journal of Thomas Moore*, ed. Wilfred S. Dowden, 6 vols (Newark: University of Delaware Press, 1983–), 1: 355: 'The feeling of natural objects, which is there expressed not caught by B. from Nature herself but from him, Wordsworth, and spoiled in the transmission.' In *The Storm-Cloud of the 19th Century* (1884), John Ruskin, by contrast, calls Byron 'the most accurate, and therefore most powerful, of all modern describers' of nature. See *The Storm-Cloud of the 19th Century* in *The Works of John Ruskin: The Library Edition*, ed. E. T. Cook and Alexander Wedderburn, 39 vols (London: George Allen, 1903–1912), 34: 7–80, 44. For Ruskin, the eruption of Krakatoa in 1883 cannot on its own adequately explain the marked and protracted deterioration of the weather (78). Curiously, he cites Byron as a witness of the benevolent weather of an earlier age (9) before plague-winds had become both the meteorological expression and divine punishment of human faithlessness.
11. Morton D. Paley, *Apocalypse and Millennium in English Romantic Poetry* (Oxford: Clarendon Press, 1999), 202.
12. See Byron's letter of 29 July 1816 to Samuel Rogers in *Byron's Letters and Journals*, ed. Leslie A. Marchand, 12 vols (London: John Murray, 1973–1982), 5: 86: 'we have had lately such stupid mists–fogs–rains—and perpetual density—that one would think Castlereagh had the foreign affairs of the kingdom of Heaven also—upon his hands'.

13. One might conjecture that the specific difference of poetry lies in the alighting on combinations that are neither wholly without basis nor altogether convincing. This is not to suggest that it is desirable for a poet to stop short of the unobjectionable, given the difficulties in circumscribing in advance the domain of the latter: an utterance can lose its recognisability as metaphor if the reception of it comes in time to ignore its initial 'semantic impertinence' (for a snobbish and presentist readership, a poem may start to falter in its title to being poetry if it cannot be disentangled from the commonplaces into which acclamation transformed its imaginative insights). The utmost 'success' of a metaphor—such as that of the *leg* of a chair and the *head* of the Church—is oblivion. See Paul Ricoeur, *The Rule of Metaphor: Multi-Disciplinary Studies in the Creation of Meaning in Language* (1975), trans. Robert Czerny with Kathleen McLaughlin and John Costello (London: Routledge & Kegan Paul, 1978), 290: 'In current usage the metaphor has been lexicalized to such an extent that it has become the proper word; by this we mean that the expression now brings its lexicalized value into discourse, with neither deviation nor reduction of deviation.' Ricoeur's example, which he borrows from Michel Le Guern, is the French word *tête* [head] in its derivation from the Latin *testa* [little pot]. The dislocation of sense through which French emerges as a discrete entity from Latin involves a breakdown of the metaphoric bond, the cement in the Tower of Babel.
14. Timothy Morton, 'Byron's Manfred and Ecocriticism' in *Palgrave Advances in Byron Studies*, ed. Jane Stabler (New York: Palgrave, 2007), 155–170, 168.
15. Jerome McGann, 'Poetry, 1780–1832' in id., *Byron and Romanticism*, ed. James Soderholm (Cambridge: Cambridge University Press, 2002), 266–287, 276.
16. Martin Heidegger, '"Only a God Can Save Us": The Spiegel Interview (1966)', trans. William J. Richardson in *Heidegger: The Man and the Thinker*, ed. Thomas Sheehan (New York: Precedent, 1981), 45–72, 57. The quietism of resignation to divine providence does not capture the sense of Heidegger's remark, given that he not only exempts thinking and poetising from the sentence he passes on action but also, following Hölderlin's 'Dichterberuf' (1800/1), does not rule out that the very absence of god may help us.
17. Compare this with Jonathan Bate, 'Living with the Weather', *Studies in Romanticism* 35.3 (1996): 431–447. Referring to what he calls Byron's 'easy yoking of politics and nature' (434), Bate argues that 'Byron may be reclaimed as a prophet of ecocide' (435). Byron's use of the past tense in 'Darkness' is enough reason to be wary of emphasising the prophetic dimension of the poem.

18. The demotion of metaphor in Byron's later poetry is discussed in Ernest J. Lovell, Jr., *Byron: The Record of a Quest: Studies in a Poet's Concept and Treatment of Nature* (Hamden, CT: Archon Books, 1966), 233: 'In *Don Juan*, to be sure, metaphors have disappeared almost altogether in favor of similes.'

CHAPTER 10

Orlando's Romantic Climate Change

Thomas H. Ford

What can Romantic literature tell us about climate change? One way eco-criticism has answered this question to date has been by reading Romantic texts as empirical records of historical changes in climate. Following Jonathan Bate's argument that the eruption of Tambora was a determining material referent of Byron's 1816 poem 'Darkness', Romanticism's atmospheric figures have often been taken literally, in a strategic practice of ecocritical demetaphorization.[1] From the haze of J.M.W. Turner's paintings to the generic mutations of Mary Shelley's *Frankenstein* (1818), Romantic artworks have been seen as registering the global climatic disturbance experienced in Western Europe as the Year Without a Summer.[2] The deranged temporalities of William Cowper's greenhouse in *The Task* (1785) have similarly been interpreted as recording the strange weather caused by an earlier eruption, that of the Icelandic volcano Laki in 1783.[3] Romantic artworks have even served as data sources for measuring changes in atmospheric composition, on the basis of a statistically significant correlation between the chromatic ratios of paintings of sunsets from 1500 to 1900 and climatic data sets reconstructed from ice cores, meteorological records and other historical sources.[4]

T. H. Ford (✉)
School of Humanities and Social Sciences, La Trobe University,
Bundoora, VIC, Australia
e-mail: t.ford@latrobe.edu.au

A. Collett, O. Murphy (eds.), *Romantic Climates*,
https://doi.org/10.1007/978-3-030-16241-2_10

The Romantic period also marked a pivotal moment in the semantic development of our climatic vocabulary, one that directly concerned the distinction between literal and metaphoric weather. Only around 1800 did the word 'atmosphere' begin to be used in ways that we would now tend to call metaphoric—as in a 'poetic atmosphere', a 'literary atmosphere' or a 'political atmosphere'. As it became more recognisably metaphoric, on the one hand, atmosphere also became more rigorously and scientifically conceptualised on the other. Historians of science date modern meteorology to around 1800. Such terminological shifts as those from 'fixed air' to 'carbonic acid gas' and from 'dephlogisticated air' to 'oxygen' indicate a comparable reorganisation of atmospheric chemistry a decade earlier. Respiratory physiology was likewise transformed in these years. But this semantic differentiation of metaphoric from literal atmospheres—of poetry becoming airy just as chemistry also was—remained unstable and only hazily uncertain in Romanticism. The discursive distinction between fictitious and factitious airs remained, in Raymond Williams's terminology, emergent, not yet having settled down into unequivocal conceptual opposition. For Williams, Western ideas of nature underwent a fundamental reconfiguration in Romanticism. There was a new conception of nature in the 'green language of Wordsworth and Clare', a 'sense of nature as a refuge, a refuge from man; a place of healing, a solace, a retreat'.[5] But there was also a scalar shift in the exploitation of nature through new extractive and technoscientific processes—a great ecosocial transformation that has more recently been described as a threshold of the Anthropocene. These two apparently opposing dynamics should be seen, Williams argued, as contradictory parts of a single social process. The emergent atmospherics of Romanticism, I claim, make this paradoxical identity peculiarly legible. Around 1800, crosscurrents of significance still shuttled between metaphoric and literal registers of atmosphere. For example, the chemist Humphry Davy proof-read William Wordsworth's second edition of *Lyrical Ballads* (1800), as atmospheric writing much more generally slipped ambiguously between observation and self-reference. As an inchoate cloud of feeling, not quite yet a structure, one that blurred metaphoric identities with literal denotation, the hazy semantics of atmosphere provided a means for remarking and giving expression to Romantic nature's radical ambiguity, this split that fell between poetic expressions and their material referents—between the human and the non-human worlds—dividing each internally from itself.

In what follows, I examine how Romanticism's climate, having migrated into the aesthetic vocabularies of literary history and criticism, was taken up in Virginia Woolf's 1928 novel *Orlando: A Biography*. Woolf's allusive reworkings of Romantic nature poetry in that text present persistent ambiguities of atmospheric signification. Her weather signs operate at once literally and materially but also metaphorically and figuratively, pointing to a zone of cloudy indeterminability between metaphoricity and literality, culture and environment. In Romantic nature lyrics, atmospheric observations doubled as poetic self-observation, so that, as M.H. Abrams argued, 'properties of the landscape became metaphors of the evolving meditation on the relation of mind to nature'.[6] Woolf's model for writing the climatic history of Romantic literature keeps these self-reflective metaphors alive, making them operational as an organising principle of literary-historical knowledge. Her citations of atmospheric Romanticism articulate ways in which it can bear meaningfully on climate change, but which become unavailable when it is taken too exclusively literally. In re-mediating Romanticism's lyric atmospheres, Woolf makes them interpretable as historiographic figures of changes in the relationship between mind and nature, between language and climate. They may even become comprehensible as figures of that relationship's transfiguration in our current climate crisis, when political and cultural atmospheres emit the literal air about us.

Midway through Woolf's comic romance of English literary history, Orlando looks out over the city of London at midnight on the final day of the eighteenth century. Like Wordsworth upon Westminster Bridge, she applies the organising gaze of the prospect poem to the metropolis, with lyrically Romantic results. As first surveyed by her, the city embodies the eighteenth century as an age of reason: 'Now—she leant out her window—all was light, order and serenity'.[7] Her vision is unobscured by any smoke or mist; the heavens are clear and still. Indeed, there is no air perceptible in her field of vision at all: 'In the extreme clearness of the atmosphere the line of every roof, the cowl of every chimney was perceptible; even the cobbles in the streets showed distinct one from another' (143). Orlando sees the orderly architecture of civilisation; the atmosphere through which she sees that architecture has disappeared transparently before its object. But even as she watches, the scene is changing. At the first stroke of midnight, a cloud appears, darkening and spreading

> with extraordinary speed. ... As the ninth, tenth and eleventh strokes struck, a huge blackness sprawled over the whole of London. With the twelfth stroke of midnight, the darkness was complete. A turbulent welter of cloud

> covered the city. All was darkness; all was doubt; all was confusion. The Eighteenth century was over; the Nineteenth had begun. (144–145)

Woolf narrates the turn from one century to the next as a literal climate change: 'A change seemed to have come over the climate of England' (146). It is at once a crossing of a meteorological threshold and a passage from one all-encompassing mentality into a new cultural field. The atmosphere becomes so laden with water that the beams of the sun are diminished, muting the colours of the world. The immediately climatic effects of this change quickly lead to broader social changes: 'Rugs appeared; beards were grown; trousers were fastened tight under the instep' (147). Walter Benjamin once argued that the ubiquitous upholstery, coverings and plush of nineteenth-century interior decoration were designed to capture and retain the traces of its bourgeois inhabitants: the philosophy of furniture as inscriptive interiority.[8] In *Orlando*, Woolf explains this same phenomenon in terms of climatic inscription: 'The chill which he felt in his legs the country gentleman soon transferred to his house; furniture was muffled; walls and tables were covered; nothing was left bare … the whole sky as it spread wide above the British Isles was nothing but a vast feather bed' (147–148). Woolf even carries the plush of the nineteenth-century interior deep into the nineteenth-century subject: 'the change did not stop at outward things. The damp struck within. Men felt the chill in their hearts; the damp in their minds. … Love, birth and death were all swaddled in a variety of fine phrases' (147). The cloud that overshadows the dawn of the nineteenth century transforms everything from geopolitics to literary sensibility: 'Thus the British Empire came into existence; and thus—for there is no stopping damp; it gets into the inkpot as it gets into the woodwork—sentences swelled, adjectives multiplied, lyrics became epics, and little trifles that had been essays a column long were now encyclopaedias in ten or twenty volumes' (147–148).

It is a joke, of course, as this image of watery ink overflowing into the hypergraphia of nineteenth-century literature makes unmistakeable. The joke works by abrupt demetaphorization: it takes the atmospheric vocabulary of Romantic literary culture literally, and so renders it ludicrous. Woolf's satiric account of nineteenth-century knowledge production as the conversion of essays into voluminous encyclopaedias, for instance, is a not-so-sly satiric allusion to John Ruskin's *Modern Painters*, first drafted in response to an attack on Turner in *Blackwood's Magazine* in 1836 and completed, 24 years later, with the publication of a fifth substantial volume. In the third volume of 1856, subtitled *Of Many Things*, Ruskin introduced into the critical lexicon the concept of 'the pathetic fallacy', the unwarranted poetic ascription of human affective meanings to non-human nature, defining it as a characteristic of what he called 'modern poetry', but which would now more usually be called Romantic poetry. That fallacy is then linked in complex ways in the following chapters *Of Many Things* to a further defining feature of Romantic art—its cloudiness. Woolf's narration of the onset of Romanticism closely tracks Ruskin's conceptual movement here in its rhetorical staging of the transition from clarity to obscurity, and also in the collapse of the scenic and essentially visual point-of-view that this transition is held to entail:

> We turn our eyes, therefore ... from these serene fields and skies of mediaeval art, to the most characteristic examples of modern landscape. And, I believe, the first thing that will strike us ... is their cloudiness. Out of the perfect light and motionless air, we find ourselves on a sudden brought under sombre skies, and into drifting wind; and, with fickle sunbeams flashing in our face, or utterly drenched with sweep of rain, we are reduced to track the changes of the shadows on the grass, or watch the rents of twilight through angry cloud. And we find that whereas all the pleasure of the mediaeval was in stability, definiteness, and luminousness, we are expected to rejoice in darkness, and triumph in mutability; to lay the foundation of happiness in things which momentarily change or fade; and to expect the utmost satisfaction and instruction from what it is impossible to arrest, and difficult to comprehend. We find, however, together with this general delight in breeze and darkness, much attention to the real form of clouds, and careful drawing of effects of mist; so that the appearance of objects, as seen through it, becomes a subject of science with us. ... So that, if a general and characteristic name were needed for modern landscape art, none better could be invented than 'the service of clouds'.[9]

For Ruskin, Romantic modernity was when atmosphere became palpable and ever-present, an inescapable dimension of social life. The Romantics linked air to aesthetic pleasure, discovering in atmospherics a source of 'general delight'. And this pleasure corresponded also to a specifically modern form of knowledge, a 'science' of atmospheric appearance. The fickleness and drift of clouds staged a new constellation of knowledge and pleasure, a figure perceptible in their mutability and transiency and in bluster and darkness. Ruskin further related the cloudiness of Romantic artworks to their anti-humanist and anti-representational experimentalism, and to their scepticism in natural religion: to the negative pleasures art discovered and cultivated in 'every aspect of age and desolation which emancipates the objects of nature from the government of men' and to its 'total absence of faith in the presence of any deity therein'.[10]

What then emerged for Ruskin through the new thematic prominence of atmosphere in Romantic art was a formal refocusing of aesthetic understanding on processes of mediation (and in particular on blockages and absences of mediation), on temporal discontinuities and on failures of comprehension and communication. Clouds figured for him the time of nature as exterior to all human government and staged the disappearance of any transcendental assurance of nature's meaningfulness. In Romantic modernity, mediation came inescapably into view as the cloudiness into which the viewing subject himself or herself also finally collapsed. But this medium, abstracted from all certain content and from all transcendent guarantees of meaning, nonetheless figured a negative human continuity with the natural world, a new aerial blending of knowledge and pleasure. For in ways later echoed by Williams and others, Ruskin identified a radical ambiguity in Romanticism's newly affective conception of nature. It gave rise to the pathetic fallacy, that peculiarly modern rhetorical form which attributes to the natural world emotional and communicative capacities generally held to pertain only to human subjects, claiming for instance that autumn leaves dance and that the sea is cruel. But equally, far from being a general instance of the pathetic fallacy for Ruskin, the cloudiness of modern cultural production in fact expressed an antipathetic interpretation of nature as finally exterior to all human systems of meaning—and so, and precisely for this reason, as forming the subject of a modern science of aesthetic appearance.

For Steven Connor, *Orlando*'s Romantic climate change is 'one of the rare attempts to offer something like a meteorological history' of modern culture.[11] Woolf's revolution in the air collapses culture into weather, 'two different orders' which for Connor can be 'connected only by mediations so vast and complex that no non-trivial determinations can really be established'.[12] In mashing together the vast intermediating series of relationships that separate weather from culture, Woolf's meteorological history comically treats these as unproblematically interchangeable terms. 'It is the very legibility of cultural-historical and meteorological parallels' in *Orlando*, Connor argues, 'that make them instances of the pathetic fallacy'.[13] This straightforward legibility is essential to Woolf's joke. She literalises the Romantic cultural history of climate change to a point of manifest absurdity, and the atmospheric materials which she submits to this satiric procedure are those provided by Ruskin himself, and by the poets—William Wordsworth, Samuel Taylor Coleridge, John Keats, Walter Scott and Alfred Tennyson most prominently—who were acclaimed by Ruskin for their 'service of the clouds', but also criticised in his historical definition of the pathetic fallacy. And yet, even as it pokes fun at the consoling idea of a natural order affectively answerable to the changing moods of human history, Woolf's 'turbulent welter of cloud' is more than simply a joke at the expense of nineteenth-century cultural cloudiness, for it shares in the productive ambiguities of the climatic literature that it also presents as its punchline. Her meteorological history of literature is 'half laughing, half serious', as she noted of the novel in which it appears.[14] To read Woolf's climate change as a single-voiced satire of its Romantic sources is to distinguish sharply between the figurative and literal meanings of its atmospheric vocabulary, respectively historical and meteorological, cultural and natural. Such an interpretation would implicitly be premised on the claim that Woolf treats a series of essentially metaphoric expressions and phrases—those of the climate of opinion and of literary and cultural atmospheres, for instance—as if they were strictly literal. As such, it would posit history and cultural meaning as not just separable from the aerial flux of weather, but as existing on the far side of some uncrossable divide from it. The comic absurdity generated by taking one thing for another—culture for atmosphere—depends on those two things being recognised as properly distinct from each other, even conceptually incommensurable. Yet this is precisely the type of unbridgeable distinction Woolf's literary atmospherics is at work throughout the novel to blur and unsettle.

Some two centuries and one gender-transition prior to this 1800 climate change, the young Orlando binds his wandering heart to an oak tree, as Woolf narrates what it feels like to fall in love with nature in the effort to write about it:

> He sighed profoundly, and flung himself—there was a passion in his movements which deserves the word—on the earth at the foot of the oak tree. He loved, beneath all this summer transiency, to feel the earth's spine beneath him; for such he took the hard root of the oak tree to be; or, for image followed image, it was the back of a great horse that he was riding; or the deck of a tumbling ship—it was anything indeed, so long as it was hard, for he felt the need of something which he could attach his floating heart to;

> the heart that tugged at his side; the heart that seemed filled with spiced and amorous gales every evening about this time when he walked out. To the oak tree he tied it and as he lay there, gradually the flutter in and about him stilled itself; the little leaves hung, the deer stopped; the pale summer clouds stayed; his limbs grew heavy on the ground; and he lay so still that by degrees the deer stepped nearer and the rooks wheeled round him and the swallows dipped and circled and the dragonflies shot past, as if all the fertility and amorous activity of a summer's evening were woven web-like about his body. (7)

So closely does it track its Romantic literary sources, Woolf's amorous pastoral might almost be read as a transcription of the greater Romantic lyric. Take the animals, for instance, which, in approaching Orlando, have wandered into these sentences from the bestiary of Romantic nature poetry. The swallows that dip and circle have veered there from lines in Keats's early fragment, 'Calidore' (1817):

> Scarce can his clear and nimble eye-sight follow
> The freaks and dartings of the black-winged swallow,
> Delighting much, to see it half at rest,
> Dip so refreshingly its wings, and breast,
> 'Gainst the smooth surface, and to mark anon,
> The widening circles into nothing gone.[15]

The deer that 'by degrees ... stepped nearer' recall Wordsworth's 'The White Doe of Rylstone' (1815) that 'Drew softly near her, and more near— / Looked round—but saw no cause for fear'; the rooks bend into ambient spirals the straight path taken by Coleridge's last homeward rook through the dusky air of 'This Lime-Tree Bower My Prison' (1800).

And the Romantic atmosphere of Woolf's paragraph is far more deeply interfused than even these poetic animal tracks suggest. The narrative sequence of these three sentences, passing from turbulent and fluttering human emotions to natural stillness and tranquillity, recapitulates the Wordsworthian model established canonically in 'Tintern Abbey' (1798).[16] The sequence begins, rearranged chronologically, with 'glad animal movements' in which affect is immediately embodied in action. But this first unreflecting stage is soon troubled by the anxiety of human impermanence and by the incursions of an unhappy awareness of temporal mutability. Orlando yearns to feel something hard and enduring, 'beneath all this summer transiency', and it is to this end that he binds his heart to the oak and falls in love with nature. This act of Romantic attachment then appears to transform temporal experience, just as promised in 'Tintern Abbey': duration dilates, time slows, the pale clouds stay, and Orlando enters something like that 'blessed mood, / In which the affections gently lead us on', until 'we are laid asleep in body, and become a living soul', adrift in an affective mesh of amorous environmental life.[17] For the poet of 'Tintern Abbey', the still life of things, glimpsed in this mood of suspended respiration, confirms the recognition of 'nature and the language of the sense' as providing.

> The anchor of my purest thoughts, the nurse,
> The guide, the guardian of my heart, and soul
> Of all my moral being. (ll.109–112)

And, as in 'Tintern Abbey', the anchoring of thought that Woolf's poet, Orlando, seeks in nature is realised, perhaps paradoxically, by language and sensation melting into air and into each other, in the breathy interchange between 'this corporeal frame', his body, and the 'summer's evening' that is said to be 'woven web-like' about it. This is, perhaps, as Walter Benjamin wrote in definition of his concept of aura, 'what it means to breathe the aura of those mountains, of that branch'—or of that oak tree.[18]

It is of course anachronistic to claim Orlando as a Romantic poet at this point in the novel: in the very next paragraph, Queen Elizabeth comes to visit. The thick weave of Romantic intertextual allusions in this paragraph conflicts with the internal schema of dates that structures the novel's telling of history. But Orlando is not just a character in this novel; she/he is also something like the mercurial spirit of English literature and, as such, seems in elusive moments of self-absence to be biographically capable of embodying the historical leaps, retrospective influences and disjunctive connections that characterise histories of a specifically literary nature, particularly those constructed within works of literature themselves. As the narrative voice states in *Orlando*, 'it is a difficult business—this timekeeping; nothing more quickly disorders it than contact with any of the arts' (200). Beneath the mutability of Orlando's changes, and possibly even motivating the character's gender instability, there remain his/her environmental Romanticism and corresponding struggle with nature writing, which are figured most paradigmatically in the oak tree to which his/her changeable heart is here permanently, textually, attached.

The experience first narrated in this paragraph on the oak tree later forms the subject of a poem, titled 'The Oak Tree', that is composed by Orlando over centuries, linking the novel together with the biographical subject it narrates, from this Elizabethan moment through to the present of the novel's final sentence, 'Thursday, the eleventh of October, Nineteen hundred and Twenty Eight' (215). Composition for Orlando signally includes publication and public consumption. Her manuscript speaks to her, telling her that it wants to be read. This inspires her for 'the first time in her life' to turn 'with violence against nature'—for only humans can read (178). But literary composition in *Orlando* also involves decomposition. As if constructing a site-specific and unobserved artwork that mixes environmental art, conceptual art and book art, Orlando finally returns to her oak to abandon her published book at its foot, letting it 'lie unburied and dishevelled on the ground'; she sees her unread poem to be 'a secret transaction' with the tree, 'a voice answering a voice' (213). In this final act of literary composition as composting, in which the book merges into decaying leaf litter, that answering voice can be understood as linguistic only negatively—as an answer that is communicated through the disappearance of language and as the material disappearance of the text. Language intersects with nature in communicative failure. The obdurate silence of linguistic materiality—of the book as an unburied but mute decaying body—stages a poetic interchange of feeling with the non-responsive alterity of the natural world, a non-voice that answers with silent mulch.

Broadly speaking, Woolf's text signals two ways of reading this oak tree, two divergent interpretations which correspond to the two sides of the radical ambiguity Raymond Williams located in Romanticism's green language. One might call them the arborescent reading and the nebulous reading, of the tree understood as a symbol, or as a mobile haze of meaning. The first would link this oak to the novel's central thematic interest in landed property, and its exploration of that property's participation in legal structures of patriarchy along with its affective value for a nativist ideology of Englishness. In a passage added to *The Prelude* in 1832, Wordsworth described Edmund Burke, the great Romantic-period theorist of modern conservatism, as an oak tree:

> I see him, old but vigorous in age,
> Stand, like an Oak whose stag-horn branches start
> Out of its leafy brow, the more to awe
> The younger brethren of the grove.[19]

The awe-inspiring inter-generational oak was Burke's own figure for the settled order of the landed aristocracy: 'You people of great families and hereditary Trusts and fortunes', he wrote to the Duke of Richmond in 1772, 'if you are what you ought to be are the great Oaks that shade a country and perpetuate your benefits from Generation to Generation'.[20] In his *Reflections on the Revolution in France* (1790), Burke returned to this image of 'the British oak' to describe traditionary constitutional settlement as an overarching political dispensation under the shadow of which its untroubled subjects reposed and quietly chewed their cud.[21] Like Orlando, heart bound to the great tree, the English had hearts of oak: solid, unchanging; ramifying out into the certainties of inherited social hierarchy; vigorous in maritime force-projection; deeply rooted in the native soil of time-honoured convention.

Williams alludes to some of the ideological dimensions of this image when he reminds us that:

> we have to remember—which we usually don't, because a successful image was imposed on us—that our first really ruthless capitalist class, taking up things and men in much the same spirit and imposing an at once profitable and pauperizing order on them, were those eighteenth-century agrarians who got themselves called an aristocracy, and who laid the real foundations, in spirit and practice (and of course themselves joining in) for the industrial capitalists who were to follow them.[22]

In the paragraph preceding Orlando's act of love, Woolf also reminds the reader of the transformative economic structures that were naturalised in this figure of the traditionary oak. Orlando surveys the surrounding countryside from an eminence, again enacting the proprietorial gaze of the eighteenth-century prospect poem. She looks out across counties and over the Channel, on which there are 'galleons setting out to sea; and armadas with puffs of smoke from which came the dull thud of cannon firing', out towards London, with its spires and smoke, and even across, 'when the wind was in the right quarter', to 'the craggy top and serrated edges of Snowdon herself'—the whole panoply of British imperial maritime expansion, its basis in the settled landscape of agrarian capitalism, and its Wordsworthian peaks of ideology laid out before him—and concludes: 'That was his father's house; that his uncle's. His aunt owned those three great turrets among the trees there. The heath was theirs, and the forest; the pheasant and the deer, the fox, the badger, and the butterfly' (6–7).

For Orlando, 'counting, gazing, recognizing', it all comes down to who owns what, even the butterfly (6). Read in these terms, attaching one's heart to an oak tree—falling in love with nature—means affectively identifying with a system of property relations and its attendant political regime, a whole world of patriarchy and profit-taking; of identifying with it, and with one's privileged place within it, and yet disavowing the true content of this identification in the same act.

What this reading struggles to account for, however, is the simultaneous semantic specificity and lability of this passage, and of the affective relations it constructs—the sense in which it presents a unique yet indefinite aggregate of mobile elements, a complex web of unstill perceptions and sensations. With primary reference to the writing of Woolf, Gilles Deleuze and Félix Guattari described such a singular linguistic assemblage with the term 'haecceity', a concept that named

> a set of speeds and slownesses between unformed particles, a set of nonsubjectified affects … assemblages that are inseparable from an hour, a season, an atmosphere, an air, a life. … Climate, wind, seasons, hour are not of another nature than the things, animals, or people that populate them, follow them, sleep and awaken within them.[23]

As image follows image, Woolf's paragraph can be read as a singular collection of just such transformations and becomings, a dynamic cloud of unformed particles and unassignable effects, rather than as the narrative of the grounding of a newly stable subject on the bedrock of private property. Orlando takes the tree's root to be the spine of the earth; then it becomes a horse, a line of flight also followed in Franz Kafka's 'Wish to become an Indian' (1913):

> If only one were an Indian, already there, and on the running horse, aslant in the air, trembling and quaking over the trembling ground, until the spurs were lost, for there were no spurs, until one threw the reins away, for there were no reins, and barely saw the land in front as a smoothly mown heath, already without the horse's neck and horse's head.[24]

The oak tree acts as comparable vehicle of abstraction and flight; it too leans transversally into the air of its own textual representation. It is a spine, a horse, a tumbling ship bearing an unknown affective freight of heart-filling gales.

Orlando looks to the oak tree as a point of existential solidity; he desires in it something 'hard', perhaps even something phallic. He believes that what he seeks there lies 'beneath all this summer transiency'. In this paragraph, he appears to behave exactly as a budding Romantic nature poet should. Late in his life, Wordsworth would recall that as a child he was often

> unable to think of external things as having external existence & I communed with all I saw as something not apart from but inherent in my own immaterial nature. Many times while going to school have I grasped at a wall or tree to recall myself from this abyss of idealism to the reality.[25]

Orlando similarly hopes to ground himself by grasping at a tree, in an effort to escape the abyss of fluttering subjectivity through contact with the fixed certainties of external reality, 'to feel the earth's spine beneath him'. But instead, the objective world appears to dissolve into the same untethered mutability experienced by his floating heart. The earth's spine proves to be strangely atmospheric, a haze of associations each at once specific and yet unceasingly slipping away into something else, at once definite—singular, unmistakably itself—but also 'anything indeed'. The 'spiced and amorous gales' that inspire Orlando's floating heart also fill the sails of the tumbling ship; they gust into the oblique air of a galloping horse. Orlando attaches his heart to the oak in order to still his 'summer transiency', but what happens instead is that he becomes only more immersed in this hour, this season, this atmosphere, this air, this life—drowsily suspended in a fluctuating world in which 'all the fertility and amorous activity of a summer's evening were woven web-like about his body'. A nature-loving heart remains a floating heart. Rather than existential stability, what transpires in this intertextual rehearsal of the lyric interaction with Romantic nature is a dissolution of thought and feeling into the strange externality of a citational atmosphere: 'everything was partly something else, as if her mind had become a forest with glades branching here and there; things came nearer, and further, and mingled and separated and made the strangest alliances and combinations in an incessant chequer of light and shade' (212).

'The spine of the earth', 'the little leaves hung', 'woven web-like about his body': Orlando's erotic encounter with the oak, these phrases suggest, is also an encounter with the spine of a book, the leaves of its pages, the woven web of its text. The dense network of borrowings throughout Woolf's novel works to the same effect: the enshrouding evocative aura of this paragraph, as we have seen, is constructed from cited textual materials and poetic allusions that constitute a typographical atmosphere, a hazy penumbra of half-heard Romantic-period echoes and resonances. What provokes Orlando's visit to the oak is a specifically literary problem, the problem of the impasse into which the literary representation of nature inevitably enters. However green your language, you can never write your way out of language and into nature, and the attempt to do so yields not nature itself but only more writing. Orlando, Woolf writes,

> was describing, as all young poets are forever describing, nature, and in order to match the shade of green precisely he looked (and here he showed more audacity than most) at the thing itself, which happened to be a laurel bush growing beneath the window. After that, of course, he could write no more. Green in nature is one thing, green in literature another. Nature and letters seem to have a natural antipathy; bring them together and they tear each other to pieces. (5)

Green in nature, this passage suggests, can be represented only negatively. Letters are not natural, and the attempt to make them so—to devise a literary language of nature, a poetically green language—is self-defeating. The word 'green' is not itself green, even when it is printed on green paper, as were 15 copies of *Orlando* in a special limited first edition. Woolf appears to pass a grim judgement here on any attempted literary mediation of the affective powers of the natural world. A poetic language—that of Romanticism, for example—that aspires to express the green of nature as it exists outside language would be at best an instance of the pathetic fallacy. In more current critical terminology, it would act as an ideological screen for the disavowal of the non-natural power relations it simultaneously encodes.

But notice that nature is doubled in Woolf's apparently unequivocal statement of the mutual exclusion of writing and nature: the antipathy between letters and nature is itself described as 'natural'. If nature has natural feelings—antipathy, for one, and so and perhaps other affective modes, other *pathē*—so too, it would seem, do letters themselves. Vaguely, in their non-identity with each other, and indeed via their mutual exclusion, nature and letters appear to overlap, as if a potential channel of communication could function in virtue of its own impossibility, a language that only came to speech by being pushed past the limits of its comprehensibility as language. In *Orlando*, falling in love with Romantic nature is presented as a complex atmospheric solution to this literary problem, that of representing nature as it is in itself. Falling in love with a web-like atmospheric texture mobilises both letters and nature so that, rather than being 'one thing' and 'another', they instead come to figure 'anything indeed', indeterminately. Atmosphere here sets into communication, however elusively, what Williams called 'elements of social and material (physical and natural) experience' that are otherwise unrepresentable and non-communicable in the available terms of a given semantic formation.[26] Language becomes green in Romanticism through its capacity for differential self-reference, as with the doubled appearance of 'nature' in Woolf's sentence, and through the modes of speculative self-reflection this enables, as when the immersive atmosphere woven about Orlando's body is disclosed as a linguistic artefact, an aerial construction of words. 'We have to remind ourselves that there is such a thing as atmosphere', Woolf wrote the year before she began *Orlando*.[27] A few years earlier, she wrote: 'It's atmosphere that makes English literature unlike any other—clouds, sunsets, fogs, exhalations, miasmas'.[28] As Jayne Lewis has observed, 'atmospheric reading' in Woolf 'never pretends to pin stable signs on some posited referent. Nor is it merely a lyric interlude. It keeps disappearing into writing, and writing itself is construed as an activity and as the atemporal physical form to which that activity gives rise'.[29] Williams attributed literature's capacity to communicate social and material experience 'in solution' to this same duality, describing the literary work of art as constituted by 'objectified conventions and notations', but noting also that 'the making of art is never itself in the past tense. It is always a formative process, within a specific present'—the same permanent present of artistic creation to which, as he elsewhere argued, Romanticism brought a newly reflective emphasis with its green language.[30] Later in the novel, much closer to the present, Orlando reads her shopping list 'in a curious stiff voice at first as if she were holding the words—boy's boots, bath salts, sardines—under a tap of many-coloured water. She watched them change as the light fell on them. Bath and boots became blunt, obtuse; sardines serrated itself like a saw' (196). In its changeable materiality and its medial liminality between text, voice and light, and between audibility and visibility, language here is also always partly something else, something non-linguistic. Writing is both a process and a form, a medium and a thing, one thing taken for another, sardines for a saw. Woolf recovers this inscriptive doubleness of atmospheric textuality from English literary history—and specifically from Romanticism—embodying it in *Orlando* in the long duration of the life of modern literature.

Woolf's atmospheric reading resists the illusion of stable signs pinned to certain referents, as Jayne Lewis has argued. It apprehends fugitive historical experience only obliquely, indirectly and negatively, so that the absence of a relationship between one thing and another—between weather and culture, for example, or between green in nature and green in letters—emerges as the condition of their potential inter-communication at the edge of semantic availability. Read atmospherically, Woolf's parody appears as ambiguously self-cancelling: it relies on a distinction between literal and figurative language that it also dissolves. Rather than being the arborescent object of pure satire, atmospheric Romanticism is in fact Woolf's primary source for this complex dynamic, this separation that is also a submerged or negative connection, half laughing, half serious. It is possible to read *Orlando* as a parody of a natural history of English literature and, specifically, of English Romantic literature, in which both natural history and Romanticism appear as pathetic fallacies in the service of the conventionally patriarchal structure of historiography.[31] But it is equally possible to glimpse in *Orlando*'s creative reworking of its atmospheric sources the methodological outlines of a natural history of Romantic literature of a very different order. For all its affective prominence in Woolf's atmospheric cultural history—indeed, precisely where it is most prominent, as the laughable cloud that descends around 1800 on England and got into everything, from writing to underpants—nature must be read as if placed under erasure, disappearing illegibly into poetry.

In *Orlando*, midnight sounds twice. The first time marks the onset of Romanticism's climate at the end of the eighteenth century, as the turbulent welter of cloud overhangs English culture and suffuses the moist depths of its inward subjects. The second time midnight sounds occurs in the novel's final sentence, now, in the moment of writing that is also the moment of reading: now, 'the twelfth stroke of midnight, Thursday, the eleventh of October, Nineteen hundred and Twenty Eight' (215). These two 12th strokes may be as historically distant as 1800 and 1928, but in this novel they are also linguistically identical, which is to say, citational. Rhymed together through citation, they bear elements of a distant historical atmosphere into the historical present. Orlando takes the oak root to be a spine, a horse, a tumbling ship, a web-like aura, so that the specificity designated by one thing and another transforms instead into the otherwise indescribable texture of a common moment, a strange and singular linguistic weave of disjunctive times and spaces. To read *Orlando*'s climatic history of English literature as parodic is to maintain a strict demarcation between the literal and figurative meanings of its vocabulary of weather, on the principle that green in nature is one thing, green in literature another. In retracing the text's dense weave of allusions and echoes of atmospheric Romanticism, I have sought to show how its weather language resists such a demarcation, and that, while no doubt playful, the novel is always something more—or less—than parody. *Orlando* re-mediates the climate of Romantic writing, making it potentially available for further philological reconstruction in our own moment, when the legibility of literary atmospheres must be again reconsidered under the manufactured skies of the Anthropocene. This is what *Orlando* tells us about what Romanticism can tell us about climate change: don't take it too literally.

Notes

1. Jonathan Bate, 'Living with the Weather', *Studies in Romanticism* 35 (1996): 431–447.
2. Michel Serres, 'Science and the humanities: the case of Turner', *SubStance* 26.2 (1997): 6–21; Gillen D'Arcy Wood, *Tambora: The Eruption that Changed the World* (Princeton: Princeton University Press, 2014).
3. Tobias Menely, '"The Present Obfuscation": Cowper's *Task* and the Time of Climate Change,' *PMLA* 127.3 (2012): 477–492.
4. C.S. Zerefos et al., 'Atmospheric effects of volcanic eruptions as seen by famous artists and depicted in their paintings', *Atmospheric Chemistry and Physics* 7.15 (2007): 4027–4042.
5. Raymond Williams, 'Ideas of Nature', in *Problems in Materialism and Culture* (London: Verso, 1982), 80.
6. M.H. Abrams, 'Structure and Style in the Greater Romantic Lyric', in *From Sensibility to Romanticism*, ed. F.W. Hilles and Harold Bloom (New York: OUP, 1965), 551.
7. Virginia Woolf, *Orlando: a Biography* (London: Hogarth Press, 1992), 144. Subsequent references to this work will appear as in-text citations.
8. Walter Benjamin, *The Arcades Project* trans. Howard Eiland and Kevin McLaughlin (Cambridge, MA: The Belknap Press of Harvard University Press, 1999), 20.
9. John Ruskin, *Works* 5, ed. E. T. Cook and Alexander Wedderburn (London: George Allen, 1904), 317–318.
10. Ibid., 319–320.
11. Steven Connor, *The Matter of Air: Science and Art of the Ethereal* (London: Reaktion, 2010), 176.
12. Ibid., 176.
13. Ibid., 177.

14. Virginia Woolf, *The Diary of Virginia Woolf, Volume Three: 1925–1930*, ed. Anne Olivier Bell (San Diego: Harcourt, Brace and Co., 1981), 168.
15. John Keats, 'Calidore: A Fragment', in *Complete Poems* ed. Jack Stillinger (Cambridge, MA: Belknap Press of Harvard University Press, 1982), 21.
16. For Woolf and Romanticism, see Karen Lawrence, *Penelope Voyages: Women and Travel in the British Literary Tradition* (Ithaca, NY: Cornell University Press, 1994), and Ellen Tremper, *'Who Lived at Alfoxton?': Virginia Woolf and English Romanticism* (London: Associated University Press, 1998).
17. William Wordsworth, 'Lines Written a Few Miles Above Tintern Abbey, on Revisiting the Banks of the Wye during a Tour, July 13, 1798', *Lyrical Ballads* (London, 1798), 116–120, ll.42–43, 46–47.
18. Walter Benjamin, 'Little History of Photography', trans. Rodney Livingstone, in *Selected Writings, Volume 2, Part 2: 1931–1934*, ed. Michael W. Jennings, Howard Eiland and Gary Smith (Cambridge, MA: The Belknap Press of Harvard University Press), 518–519.
19. William Wordsworth, *The Fourteen-Book Prelude*, ed. W. J. B. Owen (Ithaca: Cornell University Press, 1985), 151.
20. Edmund Burke, *Selected Letters of Edmund Burke*, ed. Harvey C. Mansfield, Jr. (Chicago: University of Chicago Press, 1984), 184.
21. Edmund Burke, *Reflections on the Revolution in France*, ed. Frank M. Turner (New Haven: Yale University Press, 2003), 73.
22. Williams, 'Ideas of Nature', 78–79.
23. Gilles Deleuze and Félix Guattari, *A Thousand Plateaus: Capitalism and Schizophrenia*, trans. Brian Massumi (Minneapolis: University of Minnesota Press, 1987), 262–263.
24. Franz Kafka, *Die Erzählungen* ed. Roger Hermes (Frankfurt: Fischer 2002), 7.
25. *The Fenwick Notes of William Wordsworth* ed. Jared Curtis (Tirril: Humanities-Ebooks, 2007), 124.
26. Raymond Williams, *Marxism and Literature* (Oxford: OUP, 1978), 133.
27. Virginia Woolf, 'On Being Ill,' in *The Essays of Virginia Woolf IV: 1925–1928*, ed. Andrew McNeillie (London: Hogarth Press, 1994), 326.
28. Virginia Woolf, 'A Talk About Memoirs,' in *The Essays of Virginia Woolf III: 1919–1924*, ed. Andrew McNeillie (London: Hogarth Press, 1988), 181.
29. Jayne Lewis, *Air's Appearance: Literary Atmosphere in British Fiction, 1660–1794* (Chicago: University of Chicago Press, 2012), 19.
30. Williams, *Marxism and Literature*, 129.
31. See Jane de Gay, 'Virginia Woolf's Feminist Historiography in *Orlando*', *Critical Survey* 19.1 (2007): 62–72.

CHAPTER 11

Afterword: Ghosts of 1816

Gillen D'Arcy Wood

Climate Shock: The Three Stages

I have been teaching the Romantic poets to American undergraduates for 20 years. I tell them about 'The Year Without a Summer', 1816, by the lake in Geneva. It's my favorite class of the semester, to pass along a myth almost as enthralling as *Frankenstein* itself. I prefer it larger-than-life, as befits myth. I debunk nothing, critique nothing. Poets gone wild, I tell them. Doing drugs; free love. College kids, and dropouts, on summer vacation, lucky enough to be staying next door to a rock star (Byron). Just like your summer holidays. The only difference? By summer's end, between them, this high-octane group has sketched out two totems of our cultural modernity—Frankenstein's monster and the Vampire—written the first great climate change poem ('Darkness'), the opening of *Manfred*, a further gripping installment of the best-selling poem in history ('Childe Harold's Pilgrimage'), and tossed off another smash hit, 'The Prisoner of Chillon'. And, oh yes, there's 'Mont Blanc' and 'Hymn to Intellectual Beauty'. That's your summer assignment, I say: I want world-changing cultural icons and a thousand lines of imperishable verse, on my desk, first day of the fall. My students don't see the humor in it, to be honest.

G. D. Wood (✉)
Department of English, University of Illinois, Urbana, IL, USA
e-mail: gdwood@illinois.edu

A. Collett, O. Murphy (eds.), *Romantic Climates*,
https://doi.org/10.1007/978-3-030-16241-2_11

It's worth remembering that without Claire Clairmont's lust for Byron, the famous Genevan Summer would never have happened. Mary and Percy wanted Lucerne or somewhere in Italy. The Shelleys end up loathing the city itself. It was provincial, conservative, and stuffed with tourists. But Claire prevailed, mainly because Geneva was cheap, plus, of course, the vampiric lure of Lord Byron. But by June 1816, Byron has discovered for himself what we all still marvel at today, 200 years later: the groupies next door happened to be geniuses.

The concentration of literary talent by Lake Geneva in 1816 is a freak cultural accident, just as the summer itself turned out to be a freak *climate* event. And here we've arrived at the strange, dark heart of the story. What was supposed to be a summer of literary picnics, Alpine metaphysics, and boating on the lake, turns into the dreariest wet squib of a holiday in living memory. Actually, in a millennium. The year 1816 produces the wettest, coldest summer in Geneva in 450 years of weather records. Frightening storms. Flooding in the streets. Bridges washed away. Snowpacks and glaciers encroaching down the Jura Mountains, swallowing up farmland. Crops drowned, and dead cattle floating down the river. The price of bread rises inexorably with the water level of the lake, 5, 6, 7 feet above normal. At dinner parties in Geneva, guests are asked to BYO bread. Outright famine is avoided only by the arrival of shipments of Russian grain from Odessa in November.

Mount Tambora's eruption in faraway Indonesia, the source of all the chaos, occurred ten weeks before Napoleon's defeat at Waterloo, in April 1815. By a sinister Napoleonic turn, therefore, 'The Year Without a Summer', 1816, coincided with the return of English tourists en masse to Europe post-Waterloo, who had been denied access to continental sights and delights since the Peace of Amiens in 1802. A second British invasion, but with guidebooks and fed-up valets instead of the bloody bayonets and mortar shells of Waterloo a year earlier.

Here's where the summer of 1816 myth does us a disservice. The Shelleys were, on the face of it, just another set of tourists in Geneva that year, among crowds of thousands. Not quite the penniless backpackers they were in 1814, but not exactly spreading the princely largesse, either. In this sense, the reality is even better than the myth. The supreme cognitive dissonance of that summer—the story stranger even than the one handed down to us—is of armies of tourists descending on Switzerland, literary prodigies included, just in time for an epic humanitarian disaster to unfold around them. Even through our telling 'The Year Without a

Summer' story over and over—the cold, rainy summer by the lake; Byron and the Shelleys forced indoors; the ghost story competition; *Frankenstein*!—we have never fully appreciated its layered atmospheres, its bizarre global pathways, its world-historical monster shapes, where Romantic genius meets touristic banality meets ecological catastrophe.

The slightly anodyne phrase itself, 'The Year Without a Summer', criminally understates the ecological disaster that engulfed Europe, North America, and the globe for three full years after Tambora's monster eruption. 'Years of Hell' would be more accurate. Years of a global climate system gone haywire; floods and drought; storm tracks re-routed; ocean currents gone rogue; ruined crops; rampant disease; and a dull, meager sun that seemed ready to fizzle out any moment. A death toll is difficult to calculate, but certainly tens of thousands perished from starvation and disease across the transatlantic zone, and a million worldwide. That number multiplies exponentially if we include—as we should—the impact of epidemic cholera originating from the altered disease ecology of the Bay of Bengal post-Tambora, in 1816. Try to imagine the nineteenth-century absent cholera! You can't.

But how to escape the litany-of-disaster trap in re-telling the Tambora story? How to avoid the morbid enumeration of human suffering, the piling up of heart-rending vignettes like unburied corpses clogging the canals of Bangkok, or bodies lining roadsides from Yunnan to County Clare? The question is especially urgent given that the Tamboran climate shock of 1816–1818 is the most recent, most vivid historical parallel we have to the global climate shock we are experiencing now. The temperatures of 1816–1818 went down rather than up, but the flow-on impacts of extreme weather events—from agricultural collapse to unloosed vectors of epidemic disease—are tragically comparable to today. Indian drought, Canadian wildfires, the Zika virus: take your pick from our own weird climate headlines, and there's an 1816 analogue. If the Tamboran climate shock is any guide, we are in for a rough ride, especially since there's no terminal horizon to the planet-sized chemistry experiment we call industrial modernity.

So, how to rescue 'The Year Without a Summer' from the realm of myth, and recast it as a vital climate change narrative, an historical prologue to our Late Petroleum Era of the twenty-first century and its converging ecological crises? We need a model, like the climate scientists and disaster demographers use. A model of *climate shock response* we can invoke to better grasp the enormity of the humanitarian disaster that occurred in

1816–18, and its relevance to us. Luckily, we have Byron's 'Darkness' for that, and *Frankenstein*, too. We could call it Romantic Catastrophism, or the Byronic Theory of Disaster. But the catchy name doesn't matter, as long as it helps us scale up 'The Year Without a Summer' story from an entertaining classroom anecdote of Poets Gone Wild on Lake Geneva to a world-historical tragedy of urgent relevance.

Climate shock response—Byron and Mary Shelley teach us—has three distinct stages. These 'stages' are not temporally consecutive, although time is a factor. Rather, they are class dependent—an index of economic vulnerability. In other words, all three stages of climate shock response occur simultaneously, among different social groups, according to their level of resilience. The three stages are as follows: first, political violence; second, creative sympathy; third—well, the third is a sinkhole of horror that really passes beyond description. The years 1816–18 offer a luxury of instances of all three stages of climate shock response, eerily precursive of our current climate dystopia.

The political violence stage of climate shock response embraces both seditious insurrection and the authoritarian brutality employed to repress it. In 1816–18, a direct correlation existed between the incidence and level of violence and the price of grain. Starving mobs roamed County Kildare in the spring of 1817, looting granaries and attacking food convoys bound for Dublin. Continental Europe witnessed widespread rioting and looting in market towns. Bakeries were burned down in anger at the astronomical price of bread. On another plane of social protest entirely, peasant women of Switzerland were summarily beheaded for killing their starving children. In England, Lord Liverpool's government responded to the desperation of the people with draconian force. To quell panic, it suppressed publication of agricultural quarterly reports for 1816 and suspended habeas corpus. Provincial jails filled to overflowing across the kingdom, while scores of hungry rioters were hanged or transported to the penal colonies of Australia.

Countermanding this spasmodic exercise of state violence against a starving and desperate citizenry was an equally broad but fitful humanitarian impulse I am calling 'creative sympathy'. Creative, sympathetic responses to the climate shock of 1816–18 took many forms, from the totemic art of *Frankenstein* and 'Darkness', to the *Kornvereine* in Austria—an innovative emergency government program for grain distribution—to the invention of the first bicycle prototype, to replace the tens of thousands of dead horses. Despite the deeply callous response overall from

Liverpool's government to the social emergency, the Poor Employment Act of 1817—which introduced the novel policy idea of public works programs to alleviate economic distress—was an act of creative sympathy. Likewise, the publication of the first modern weather map—by Heinrich Brandes in 1820—was a creative sympathetic response to the climate shock of that year.

So was the innocuous-sounding Dublin Association for the Suppression of Mendicity (street begging)—organized by a group of middle-class Irish doctors who kept meticulous records both of the weather and of the sick, starving, and homeless flocking to Ireland's cities. These activist medical men harassed the Irish secretary with pamphlets and street theater protests and wrote letters to Parliament warning of the imminent collapse of the Irish Kingdom. Robert Peel, Byron's old classmate at Harrow and Chief Secretary for Ireland at the time, proved heartlessly obdurate during the crisis, even when a drowned husk of green corn arrived in the mail, addressed to him, at Dublin Castle. But Peel was not impervious to climate shock. Something inside his Tory soul stirred and then rebelled against the entrenched modes of laissez-faire administration. In the wake of the typhus epidemic that swept Ireland in 1817–1818, he quietly empaneled a committee to study disease and the conditions of the poor. This humble committee evolved, over decades, to become the first British Board of Health. Again, creative sympathy.

Reaction to the shock experience of climate change—and its terrible human costs—is often violent and shatteringly inhumane. The Tamboran case study offers innumerable demoralizing examples. But there is this other strain to 'The Years Without a Summer', a symphonic, finer tone. Out of the climate disaster of 1816–1818, we can also trace seismic soul-searching, epic acts of charity and self-sacrifice, as well as rapid-fire bursts of artistic, technological, and administrative innovation. In rural Ireland—abandoned by the state—priests formed makeshift hospitals and died by the dozens at the bedsides of fever victims. In Northern Germany, private co-operatives, led by the affluent elite, bypassed sluggish authorities to import grain directly from Russia to feed the starving in their communities. And at the Villa Diodati, of course, Byron wrote 'Darkness', a masterpiece of climate affliction buried, like an Etruscan vase, for 200 years, to be excavated and raised to the light in this, our brand new era of climate shock.

In these first two stages of climate shock response—violence and sympathy—the social lens is restricted to the social elite, the bourgeoisie, and urban working classes. It was the urban working class who staged food riots and held magistrates hostage. It was the professional bourgeoisie who formed new charitable organizations and fought against a state-controlled press to publicize the crisis. And it was the ruling elite who enacted legislation and wrote poems. But to publish a letter attacking the government, smash a baker's window, or write a poem, requires energy, which requires food. And food was *scarce*.

The worst of what happened in 1816, 1817, and 1818 unfolded more quietly, more heartbreakingly, away from the towns and cities, in the vast rural hinterlands of the world, to people left with no food whatever. It is a fact of social history that a subsistence crisis fosters political violence and uproar, while a true famine unfolds in eerie silence. Geneva, home to the Byrons and Shelleys in 1816, experienced food riots the following spring, but never starvation conditions, whereas in rural communities across Europe, Asia, and in parts of North America, desperate countryfolk left their homes and everything they had, and took to the highways. The dead silence of the disaster was broken by only the shuffling of a million shoeless feet. Such was the third and final stage of climate shock response in 1816–18: the long march of the world's peasantry into death.

1816 Redivivus: Romantic Teleconnections

The relative vulnerability of human communities to the rapid climate change now unfolding across the globe is to a significant degree historically determined. For half a millennium northern countries have displaced the biophysical impact of their development through an 'extractivist' agro-industrial regime imposed on the resource-rich south, with dire consequences for both the human populations and ecosystems of those regions. These overlapping, centuries-old histories of colonialism and economic globalization now shape the inequitable distribution of human costs of a deteriorating global climate whose uneven ripple effects will deliver the worst impacts of drought, flood, and rising sea levels first to the most historically impoverished regions of the world, such as sub-Saharan Africa and the Pacific Islands. An essential characteristic of twenty-first-century climate change is thus its tendency to reinforce the asymmetric economic relationship between the industrialized north and global south.

In the coming decades, however, these historical borders and hemispheric inequities will become increasingly blurred as countries in all latitudes face critical disruptions to infrastructure, water resources, and food systems. In this sense, the increasing devastation of climate change, which has been called market capitalism's greatest failure, also represents the ironic end game of the postcolonial era: the north created climate change as a byproduct of its industrialization, but cannot meet the challenge of mitigating destructive warming absent close partnership with its rapidly developing former dominions in the global south—such as India and Brazil—who are in turn suspicious of Western-imposed constraints on their carbon-driven growth. The atmosphere is a unique planetary commons, less administrable even than the oceans. Who will regulate its runaway use as a carbon sink? Adding to the political paralysis, progressives in the developed world appear stricken by the 'wicked' character of the climate change problem, which sets two cherished liberal goals—economic development of the south and environmental protection—at seemingly irreconcilable odds. Carbon, after all, has been earth's greatest gift to the poor.

For scholars in the academic humanities, the foregoing suggests serious limits to the capacity of postcolonial and globalization models to address the systemic complexity, novelty, and planetary scope of climate change as an historical phenomenon. Just as European economic growth and colonization redrew the map of the world between 1500 and 1950, so climate change, and its attendant natural resource crises, will redraw the map of the twenty-first century and beyond. Governments will fall, social systems collapse, and environmental refugees stream across borders to populate the shrinking well-resourced regions of the planet. Postcolonialism, as a critical account of the 1500–1950 era, never represented a truly global paradigm, tied as it was to a Eurocentric core-periphery model based on the historical extension of North Atlantic trade, conquest, and settlement. The emerging map of global climate change in the twenty-first century, by contrast, exhibits a more dense, complex, and multivalent network of physical and political relations, governed by a principle of teleconnection linking remote places, peoples, and events.

The epic Pakistani floods and Russian wildfires at the outset of the 2010s, for example—byproducts of the same rogue weather system over the Asian continent—exemplify teleconnection both in their complex feedback relationship to each other and in their influence on events farther afield.[1] The historic floods of 2010 undermined the already fractured political establishment in Pakistan, while Russia responded to the devastation of its farm belt by closing down its export coffers. Twenty percent of grain stocks disappeared overnight from world markets, sending prices skyrocketing. Six months later, the high price of bread sparked protests across North Africa, culminating in the so-called Arab Spring that saw the fall of decades-old authoritarian regimes.

As this shorthand example shows, teleconnection characterizes global politics within an integrated twenty-first-century world political economy in ways both analogous to and materially dependent upon the physical operations of the newly unstable climate system. The changing climate and the geo-political sphere it increasingly impacts exhibit analogous susceptibilities to a quasi-chaotic causal physics of feedback loops, tipping points, and non-linear change. The teleconnected geo-politics of climate change demand that we speak of the Moscow bourgeois, the Pakistani peasant, and the Egyptian revolutionary in the same breath.

What are the implications of a deteriorating climate in a teleconnected globe to humanist scholars? More particular to this volume, what of the peculiar historical status of the Romantic era situated at the biophysical onset of the industrial Anthropocene? In one sense, it must seem that the luck of the Romanticist will never run out. Gifted with texts and a compressed dynamic of world-historical events central variously to New Criticism, Deconstruction, New Historicism, and Postcolonialism, scholars of Romanticism now appear indispensable to the evolution of the heretofore marginal Ecocriticism into a full-fledged historical account of the modern Anthropocenic era and its discourses. Postmodern thought instructs us to be wary of origins, periodizations, and teleological thinking. The scientific literature of global warming, however, suggests an alternate, but complementary view. If we are to embrace the non-ontological human, and acknowledge our outsized place in the material history of the earth, the origins of our global carbon-based modernity lie in the late eighteenth century. As writers and teachers, Romanticists now find themselves enabled (even impelled) to represent themselves as scholars of the Anthropocene's birth.

This entails a new globalism in Romantic studies and a new poetics of climate. Modern liberal historiography in important respects originated as a counter-discourse to the climate determinism of the eighteenth century. Postcolonialism, among others, inherited this constitutive prejudice against environmentalist accounts of social history. But 'history' has a way of eroding all certainties, and so it is with the anti-climatic anthropology of the modern disciplinary humanities. A new global study in Romanticism, modernized for the age of climate change, adopts the non-linear, teleconnected character of the climate system as the model for a new global historiography, one attentive but not tethered to the discourse and material infrastructures of European colonialism.

As a privileged literary field coincident with the rise of the Anthropocenic era, a robust Romantic eco-historicism, embodied in volumes like *Romantic Climates*, is central to the greater collective cause of climate salvation. With public discourse on the global climate emergency paralyzed by corporate-funded fake news, short-term electoral politics, and economic nationalism, Romanticist scholars of the Anthropocene are bound to enter the fray to combat its dominant ecocidal logic. The price of failure, after all, is too dreadful to contemplate—full of darkness, monsters, and other ghosts of 1816.

Note

1. An 'extraordinarily strong and prolonged extratropical atmospheric blocking event,' consistent with global warming, paralyzed the northern jet stream over Eastern Europe in July 2010, channeling heat from North Africa to Russia, while sucking monsoonal rains to the north over Pakistan. William K. M. Lau and Kyu-Myong Kim, 'The 2010 Pakistan Flood and Russian Heat Wave: Teleconnection of Hydrometeorologic Extremes,' *Journal of Hydrometeorology* 13 (2012): 392–403.

Works Cited

[Untitled News Report]. *The Times* (12 October 1816): 3.

Abercromby, Ralph, et al. *The Eruption of Krakatoa and Subsequent Phenomena: Report of the Krakatoa Committee of the Royal Society.* Ed. G.J. Symons. London: Trübner, 1888.

Abrams, M.H. "Structure and Style in the Greater Romantic Lyric." In *From Sensibility to Romanticism*, ed. F.W. Hilles and Harold Bloom, 527–560. New York: Oxford University Press, 1965.

Aristotle. *Poetics.* Trans. W.H. Fyfe. London: Heinemann, 1927.

Austen, Jane. *Emma* (1815). Ed. James Kinsley and Adela Pinch. Oxford: Oxford University Press, 2003.

———. *Persuasion* (1818). Ed. Janet Todd and Antje Blank. Cambridge: Cambridge University Press, 2006.

———. *Jane Austen's Letters.* Ed. Deirdre Le Faye, 4th ed. Oxford: Oxford University Press, 2011.

Bate, J. *Romantic Ecology: Wordsworth and the Environmental Tradition.* London and New York: Routledge, 1991.

———. "Living with the Weather." *Studies in Romanticism* 35, no. 3 (1996): 431–437.

———. *The Song of the Earth.* London: Picador, 2000.

Beattie, Andrew. *The Alps: A Cultural History.* New York: Oxford University Press, 2006.

Beaumont, M. *Nightwalking: A Nocturnal History of London.* London: Verso, 2016.

Benjamin, Walter. *The Arcades Project.* Trans. Howard Eiland and Kevin McLaughlin. Cambridge, MA: Belknap Press, 1999.

A. Collett, O. Murphy (eds.), *Romantic Climates*,
https://doi.org/10.1007/978-3-030-16241-2

———. *Selected Writings, Volume 2, Part 2: 1931–1934.* Ed. Michael W. Jennings, Howard Eiland, and Gary Smith. Cambridge, MA: Belknap Press, 2005.

Bewell, Alan. "Keats's Realm of Flora." *Studies in Romanticism* 31, no. 1 (1992): 71–98.

Black, Max. "Metaphor." In *Models and Metaphors: Studies in Language and Philosophy*, 25–47. Ithaca, NY: Cornell University Press, 1962.

Blake, William. *William Blake: The Complete Illuminated Books.* Ed. David Bindman. London: Thames & Hudson, 2000.

Boast, Rachael. "Silent Sea." Guardian, 29 May 2015.

Bohls, Elizabeth A. *Slavery and the Politics of Place: Representing the Colonial Caribbean 1770–1883.* Cambridge: Cambridge University Press, 2014.

Borodale, Sean. "Scratching for Metaphor in the Somerset Coalfields." Guardian, 18 May 2015.

Briffa, K.R., and P.D. Jones. "The Climate of Europe during the 1810s with Special Reference to 1816." In *The Year Without a Summer? World Climate in 1816*, ed. C.R. Harington, 372–391. Ottawa: Canadian Museum of Nature, 1992.

British Review. Review of *Manfred*. 1817.

Büchner, Georg. *Woyzeck* (1879). *Danton's Death*; *Leonce and Lena*; *Woyzeck*. Trans. Victor Price, 105–132. Oxford: Oxford University Press, 1971.

Bugg, John. *Five Long Winters: The Trials of British Romanticism.* Stanford: Stanford University Press, 2014.

Burke, Edmund. *On the Origin of Our Ideas of the Sublime and Beautiful*, 5th ed. London: J. Dodsley, 1767.

———. *Thoughts and Details on Scarcity, Originally Presented to the Right Hon. William Pitt in the Month of November, 1795.* London: F. and C. Rivington, 1800.

———. *Selected Letters of Edmund Burke.* Ed. Harvey C. Mansfield, Jr. Chicago: University of Chicago Press, 1984.

———. *Reflections on the Revolution in France* (1790). Ed. Frank M. Turner. New Haven: Yale University Press, 2003.

Burns, R. *Poems, Chiefly in the Scottish Dialect.* Kilmarnock: John Wilson, 1786.

Butler, Marilyn. *Romantics, Rebels, and Reactionaries.* Oxford: Oxford University Press, 1981.

———. *Mapping Mythologies: Countercurrents in Eighteenth-Century British Poetry and Cultural History.* Cambridge: Cambridge University Press, 2015.

Byron, George Gordon. *The Works of Lord Byron*, 14 vols. Vol. XI. London: John Murray, 1832.

———. *The Works of Lord Byron. Letters and Journals.* Vol. II. Ed. Rowland Prothero. London: John Murray, 1898.

———. *His Very Self and Voice: Collected Conversations of Lord Byron.* Ed. Ernest Lovell, Jr. New York: Macmillan, 1954.

———. *The Poetical Works of Lord Byron.* London: Oxford University Press, 1957.

———. *Byron's Letters and Journals*, 12 vols. Ed. Leslie A. Marchand. London: John Murray, 1973–1982.

———. *The Complete Poetical Works*, 7 vols. Vols. 1 and 4. Ed. Jerome J. McGann. Oxford; New York: Clarendon Press; Oxford University Press, 1980.

———. *The Complete Miscellaneous Prose*. Ed. Andrew Nicholson. Oxford: Clarendon Press, 1991.

Campbell, Shannon E. "Apples and Apple-Blossom Time (Wherein Jane Austen's Reputation for Meticulous Observation Is Vindicated)." *Persuasions: The Jane Austen Journal* 29 (2007): 89–98.

Chambers, Robert. Vestiges of the Natural History of Creation [1844]. New York: Wiley and Putnam, 1845.

Chandler, J. *Wordsworth's Second Nature*. Chicago and London: University of Chicago, 1984.

Clark, Timothy. *Ecocriticism on the Edge: The Anthropocene as a Threshold Concept*. London: Bloomsbury, 2015.

Clarke, Gillian. "Cantre'r Gwaelod." *Guardian*, 19 May 2015.

Clubbe, John, Ernest Giddey, and John Cam Hobhouse. *Byron et la Suisse: Deux Études*. Genève: Droz, 1982.

Cochran, Peter, ed. *Manfred: An Edition of Byron's Manuscripts and a Collection of Essays*. Newcastle upon Tyne: Cambridge Scholars Publishing, 2015.

Cohen, Jean. *Structure du langage poètique*. Paris: Flammarion, 1966.

Connor, Steven. *The Matter of Air: Science and Art of the Ethereal*. London: Reaktion, 2010.

Cox, Jeffrey N. *Poetry and Politics in the Cockney School: Keats, Shelley, Hunt and Their Circle*. Cambridge: Cambridge University Press, 1998.

Coxe, William, and Louis François Élisabeth Ramond de Carbonnières. *Travels in Switzerland, and in the Country of the Grisons*, 3rd ed., 3 vols. Vol. 1. Basle: James Decker, 1802.

Daston, Lorraine. "On Scientific Observation." *Isis* 99, no. 1 (2008): 97–110.

Davy, Humphry. *Humphry Davy on Geology: The 1805 Lectures for the General Audience*. Ed. Robert Siegfried and Robert Dott. Madison: University of Wisconsin Press, 1980.

de Almeida, Hermione. *Romantic Medicine and John Keats*. New York and Oxford: Oxford University Press, 1991.

de Gay, Jane. "Virginia Woolf's Feminist Historiography in *Orlando*." *Critical Survey* 19, no. 1 (2007): 62–72.

Deleuze, Gilles, and Félix Guattari. *A Thousand Plateaus: Capitalism and Schizophrenia*. Trans. Brian Massumi. Minneapolis: University of Minnesota Press, 1987.

Desmond, Adrian. *The Politics of Evolution: Morphology, Medicine, and Reform in Radical London*. Chicago: The University of Chicago Press, 1989.

Dimock, Wai Chee. "A Theory of Resonance." *PMLA* 112, no. 5 (October 1997): 1060–1071.

Dodds, Joseph. *Psychoanalysis and Ecology at the Edge of Chaos.* London: Routledge, 2011.

Eagleton, T. *How to Read a Poem.* Oxford: Blackwell, 2007.

Erskine, T. *A View of the Causes and Consequences of the Present War with France.* London: J. Debrett, 1797.

European Magazine. Review of *Manfred.* August 1817.

Evans, Gareth. "Poison Wine—John Keats and the Botanic Pharmacy." *Keats-Shelley Review* 16 (2002): 31–55.

Fairer, D. *Organising Poetry: The Coleridge Circle, 1790–1798.* Oxford: Oxford University Press, 2009.

———. "'Where Fuming Trees Refresh the Thirsty Air': The World of Eco-Georgic." *Studies in Eighteenth-Century Culture* 40 (2011): 201–218.

Favret, Mary A. "Everyday War." *ELH* 72, no. 3 (2005): 605–633.

Faye, Deirdre Le. *Jane Austen: A Family Record.* Cambridge: Cambridge University Press, 2004.

Felski, Rita. *The Limits of Critique.* Chicago: The University of Chicago Press, 2015.

Fermanis, Porscha. *John Keats and the Ideas of the Enlightenment.* Edinburgh: Edinburgh University Press, 2009.

Fletcher, Loraine. "Time and Mourning in *Persuasion.*" *Women's Writing* 5, no. 1 (1998): 81–90.

Fludernik, Monica. "The Linguistic Illusion of Alterity: The Free Indirect as Paradigm of Discourse Representation." *Diacritics* 25, no. 4 (1995): 89–115.

Franklin, Caroline. *The Female Romantics: Nineteenth-Century Women Novelists and Byronism.* New York: Routledge, 2013.

Frend, W. *Scarcity of Bread.* London: J. Smith, 1795.

Galperin, William. "'Describing What Never Happened': Jane Austen and the History of Missed Opportunities." *ELH* 73, no. 2 (2006): 355–382.

———. *The Historical Austen.* Philadelphia: University of Pennsylvania Press, 2013.

Gay, Penny. "The Romanticism of *Persuasion.*" *Sydney Studies in English* 5 (1979–1980): 15–30.

The Gentleman's Magazine. Review of *Manfred.* 1817.

Gerrald, J. *A Convention the Only Means of Saving Us from Ruin.* London: D. I. Eaton, 1793.

Gidal, Eric. "'O Happy Earth! Reality of Heaven!': Melancholy and Utopia in Romantic Climatology." *Journal for Early Modern Cultural Studies* 8, no. 2 (2008): 74–101.

Gill, S. "'Adventures on Salisbury Plain' and Wordsworth's Poetry of Protest 1795–1797." *Studies in Romanticism* 11 (Winter 1972): 48–65.

Glantz, Michael H., Richard W. Katz, and Neville Nichols, eds. *Teleconnections Linking Worldwide Climate Anomalies.* Cambridge: Cambridge University Press, 1991.

Goellnicht, Donald C. *The Poet-Physician: Keats and Medical Science*. Pittsburgh: University of Pittsburgh Press, 1984.
Gottlieb, Evan. "Fighting Words: Representing the Napoleonic Wars in the Poetry of Hemans and Barbauld." *European Romantic Review* 20, no. 3 (2009): 327–343.
Gravil, R. *Wordsworth's Bardic Vocation, 1787–1842*. Houndmills, Basingstoke, Hampshire: Palgrave Macmillan, 2003.
Gray, T. *Thomas Gray and William Collins: Poetical Works*. Ed. R. Lonsdale. Oxford: Oxford University Press, 1977.
Guillory, John. "Genesis of the Media Concept." *Critical Inquiry* 36 (Winter 2010): 321–362.
Hall, Dewey W., ed. *Romantic Ecocriticism: Origins and Legacies*. Lanham, MD: Lexington Books, 2016.
Harington, C.R., ed. *The Year Without a Summer? World Climate in 1816*. Ottawa: Canadian Museum of Nature, 1992.
Harkin, Maureen. "Matthew Lewis's *Journal of a West India Proprietor*: Surveillance and Space on the Plantation." *Nineteenth Century Contexts* 24, no. 2 (2002): 139–150.
Harris, Jocelyn. *A Revolution Almost Beyond Expression*. Newark: University of Delaware University Press, 2007.
Hayles, N. Katherine. *Chaos Bound: Orderly Disorder in Contemporary Literature and Science*. Ithaca: Cornell University Press, 1990.
Hazlitt, W. *William Hazlitt: Selected Writings*. Ed. J. Cook. Oxford: Oxford University Press, 2009.
Heidegger, Martin. "'Only a God Can Save Us': The Spiegel Interview (1966)." In *Heidegger: The Man and the Thinker*. Trans. William J. Richardson. Ed. Thomas Sheehan, 45–72. New York: Precedent, 1981.
Henderson, Andrea. "Mastery and Melancholy in Suburbia." *Eighteenth Century: Theory and Interpretation* 50, nos. 2–3 (2009): 221–244.
Heringman, Noah. *Romantic Rocks, Aesthetic Geology*. Ithaca and London: Cornell University Press, 2004.
Higgins, David. *British Romanticism, Climate Change, and The Anthropocene*. London: Palgrave Macmillan/Springer, 2017.
Hobhouse, John Cam. *Diaries of John Cam Hobhouse*. British Library Add. MS.56536, 1816.
Holmes, Richard. *The Age of Wonder: How the Romantic Generation Discovered the Beauty and Terror of Science*. London: HarperPress, 2008.
Horova, Mirka. "Byron and Catastrophism." In *Byron's Religions*, ed. Peter Cochran, 253–261. Newcastle upon Tyne: Cambridge Scholars Publishing, 2011.
Howard, Luke. *On the Modifications of Clouds, and on The Principles of Their Production, Suspension, and Destruction: Being the Substance of an Essay Read Before the Askesian Society in the Session 1802–3*. London: J. Taylor, 1803.

———. *The Climate of London: Deduced from Meteorological Observations, Made at Different Places in the Neighbourhood of the Metropolis*, 2 vols. London: W. Phillips, 1818–1820.

———. *Essay on the Modifications of Clouds*, 3rd ed. London: John Churchill & Sons, 1865.

Hutton, James. *Theory of the Earth, with Proofs and Illustrations.* Vol. 2. Edinburgh: Messrs Cadell, Junior, and Davies, 1795.

Huxley, Aldous. *Do What You Will.* London: Watts & Co., 1937.

Jeffrey, Francis. Rev. of *Manfred. Edinburgh Review*, September 1817, XXVIII.

Johnson, Claudia L. *Jane Austen's Cults and Cultures.* Chicago: The University of Chicago Press, 2012.

Jones, Elizabeth. "Keats in the Suburbs." *Keats-Shelley Journal* 45 (1996): 23–43.

Jones, Stephen E. "Politics and Satire." In *The Oxford Handbook of Percy Bysshe Shelley*, ed. Michael O'Neill and Anthony Howe, 148–162. Oxford: Oxford University Press, 2012.

Kafka, Franz. *Die Erzählungen.* Ed. Roger Hermes. Frankfurt: Fischer, 2002.

Keats, John. *The Letters of John Keats, 1814–1821*, 2 vols. Ed. Hyder Edward Rollins. Cambridge: Cambridge University Press, 1958.

———. *Complete Poems.* Ed. Jack Stillinger. Cambridge: Belknap Press, 1982.

———. *The Major Works.* Ed. Elizabeth Cook. Oxford and New York: Oxford University Press, 2008.

Kenyon-Jones, Christine. "'When This World Shall Be Former': Catastrophism as Imaginative Theory for the Younger Romantics." *Romanticism on the Net* (RON): 24 (November 2001).

———. "'He is a Rogue of Course, But a Civil One': John Murray, Jane Austen, and Lord Byron. " *Persuasions* 36 (2014): 239–254.

Kington, John. "Weather Patterns over Europe in 1816." In *The Year Without a Summer? World Climate in 1816*, ed. C.R. Harington, 358–371. Ottawa: Canadian Museum of Nature, 1992.

Kitson, P.J., ed. *Placing and Displacing Romanticism.* Aldershot: Ashgate, 2001.

Klingaman, William K., and Nicholas P. Klingaman. *The Year Without Summer: 1816 and the Volcano That Darkened the World and Changed History.* New York: St Martin's Press, 2013.

Lake District Search and Mountain Rescue Association. *Mountain Accidents 2012.* Cumbria: Lake District Search and Mountain Rescue Association, 2012. https://www.ldsamra.org.uk/documents/LDSAMRAAnnualReport2012.pdf/.

———. *Mountain Accidents 2015.* Cumbria: Lake District Search and Mountain Rescue Association, 2015. https://www.ldsamra.org.uk/documents/LDSAMRAAnnualReport2015.pdf/.

Lau, William K.M., and Kyu-Myong Kim, "The 2010 Pakistan Flood and Russian Heat Wave: Teleconnection of Hydrometeorologic Extremes." *Journal of Hydrometeorology* 13 (2012): 392–403.

Lawrence, Karen. *Penelope Voyages: Women and Travel in the British Literary Tradition*. Ithaca, NY: Cornell University Press, 1994.

Le Breton, Anna Letitia. *Memoirs of Mrs. Barbauld, including Letters and Notices of Her Family and Friends*. London: George Bell, 1874.

Leask, N. *Robert Burns and Pastoral: Poetry and Improvement in Late Eighteenth-Century Scotland*. Oxford: Oxford University Press, 2010.

Levinson, Marjorie. *Wordsworth's Great Period Poems*. Cambridge: Cambridge University Press, 1986.

———. *Keats's Life of Allegory: The Origins of a Style*. Oxford: Basil Blackwell, 1988.

Lewis, Matthew. *Journal of a West India Proprietor: Kept During a Residence in the Island of Jamaica* (1834). Oxford: Oxford University Press, 1999.

———. *The Monk: A Romance* (1796). Ed. D.L. Macdonald and Kathleen Scherf. Toronto: Broadview, 2004.

Lewis, Jayne. *Air's Appearance: Literary Atmosphere in British Fiction, 1660–1794*. Chicago: University of Chicago Press, 2012.

Lickbarrow, Isabella. *Collected Poems*. Ed. Constance Parrish. Grasmere: The Wordsworth Trust, 2004.

Litz, A. Walton. "*Persuasion*: Forms of Estrangement." In *Jane Austen: Bicentenary Essays*, ed. John Halperin, 221–232. Cambridge: Cambridge University Press, 1975.

Liu, A. *Wordsworth: The Sense of History*. Stanford: Stanford University Press, 1989.

Longley, Michael. "Storm." *Guardian*, 22 May 2015.

Lovell, Ernest J., Jr. *Byron: The Record of a Quest: Studies in a Poet's Concept and Treatment of Nature*. Hamden, CT: Archon Books, 1966.

Lynch, Deidre. Introduction to *Persuasion* by Jane Austen. Ed. James Kinsley, vii–xxxiii. Oxford: Oxford University Press, 2004.

Macdonald, D.L. "Pre-Romantic and Romantic Abolitionism: Cowper and Blake." *European Romantic Review* 4, no. 2 (1994): 163–182.

———. "The Isle of Devils: The Jamaican Journal of M.G. Lewis." In *Romanticism and Colonialism: Writing and Empire, 1780–1830*, ed. Tim Fulford and Peter J. Kitson, 189–205. Cambridge: Cambridge University Press, 1998.

———. *Monk Lewis: A Critical Biography*. Toronto: University of Toronto Press, 2000.

Macdonald, D.L., and Kathleen Scharf. "Introduction." In *Frankenstein; or, The Modern Prometheus: The 1818 Version*, ed. Mary Wollstonecraft Shelley, 2nd ed., 11–38. Peterborough, ON: Broadview Press, 2005.

Macksey, Richard. "'To Autumn' and the Music of Mortality: 'Pure Rhetoric of a Language without Words'." *Romanticism and Language*, ed. Arden Reed, 263–308. Ithaca: Cornell University Press, 1984.

Markley, Robert. "The Amherst Embassy in the Shadow of Tambora: Climate and Culture, 1816." In *Writing China: Essays on the Amherst Embassy (1816) and Sino-British Cultural Relations*, ed. Peter J. Kitson and Robert Markley, 83–104. Woodbridge: Boydell and Brewer, 2016.

Matthews, G.M., ed. *Keats: The Critical Heritage*. London: Routledge and Kegan Paul, 1971.
McGann, Jerome. *The Romantic Ideology: A Critical Study*. Chicago: University of Chicago Press, 1983.
———. "Byron and Wordsworth." In *Byron and Romanticism*, 173–204. Cambridge: Cambridge University Press, 2002a.
———. "Poetry, 1780–1832." In *Byron and Romanticism*, ed. James Soderholm, 266–287. Cambridge: Cambridge University Press, 2002b.
"Medical and Philosophical Intelligence." *London Medical and Physical Journal* 34, no. 3 (September 1815): 258.
Menely, Tobias. "'The Present Obfuscation': Cowper's *Task* and the Time of Climate Change." *PMLA* 127, no. 3 (2012): 477–492.
———. "Late Holocene Poetics: Genre and Geohistory in Beachy Head." *European Romantic Review* 28, no. 3 (2017): 307–314.
Moe, Melina. "Multiple Modernities in *Pride and Prejudice*." *ELH* 83, no. 4 (2016): 1075–1103.
Moore, Thomas. *The Journal of Thomas Moore*, 6 vols. Ed. Wilfred S. Dowden. Newark: University of Delaware Press, 1983–1990.
More, Hannah. *Village Politics*, 3rd ed. London: F. and C. Rivington, 1793.
Morton, Timothy. "Byron's Manfred and Ecocriticism." In *Palgrave Advances in Byron Studies*, ed. Jane Stabler, 155–170. New York: Palgrave, 2007a.
———. *Ecology without Nature: Rethinking Environmental Aesthetics*. Cambridge, MA: Harvard University Press, 2007b.
———. *The Ecological Thought*. Cambridge, MA: Harvard University Press, 2010.
———. *Dark Ecology: For a Logic of Future Coexistence*. New York: Columbia University Press, 2016.
Mountain Accidents 2012. Cumbria: Lake District Search and Mountain Rescue Association, 2012. https://www.ldsamra.org.uk/documents/LDSAMRAAnnual Report2012.pdf/.*Mountain Accidents 2015*. Cumbria: Lake District Search and Mountain Rescue Association, 2015. https://www.ldsamra.org.uk/documents/ LDSAMRAAnnualReport2015.pdf.
Nandrea, Lorri G. "Difference and Repetition in Austen's *Persuasion*." *Studies in the Novel* 39, no. 1 (2007): 48–64.
Nicolson, Marjorie Hope. *Mountain Gloom and Mountain Glory: The Development of the Aesthetics of the Infinite*. Seattle: University of Washington Press, 1997.
Nisbet, Euan. "In Retrospect." *Nature* 388, no. 6638 (1997): 137.
O'Connor, Ralph. "Byron's Afterlife and the Emancipation of Geology." In *Liberty and Poetic Licence: New Essays on Byron*, ed. Bernard Beatty, Tony Howe, and Charles E. Robinson. Liverpool: Liverpool University Press, 2008a.
———. *The Earth on Show: Fossils and the Poetics of Popular Science, 1802–1856*. Chicago: University of Chicago Press, 2008b.
Oerlemans, O. *Romanticism and the Materiality of Nature*. Toronto: Toronto University Press, 2004.

Orwell, George. *Animal Farm: A Fairy Story* (1945). Ringwood, Victoria: Penguin Books, 1964.

Paley, William. *Reasons for Contentment; Addressed to the Labouring Part of the British Public.* Stirling: F. Jolie, 1792.

———. *Natural Theology: Or, Evidences of the Existence and Attributes of the Deity, Collected from the Appearances of Nature.* London: J. Faulder, 1802.

Paley, Morton D. *Apocalypse and Millennium in English Romantic Poetry.* Oxford: Clarendon Press, 1999.

Paterson, Don. "Nostalgia." *Guardian*, 3 June 2015.

Peck, Louis. *A Life of Matthew G. Lewis.* Cambridge: Harvard University Press, 1961.

Pedgley, D.E. "Luke Howard and His Clouds." *Weather* 58 (2003): 51–55.

Pliny the Younger. *Complete Letters.* Trans. P.G. Walsh. Oxford: Oxford University Press, 2006.

Poovey, Mary. "From Politics to Silence: Jane Austen's Nonreferential Aesthetic." In *Companion to Jane Austen*, 249–260. Oxford: Blackwell, 2009.

Porter, Theodore M. *Trust in Numbers: The Pursuit of Objectivity in Science and Public Life.* Princeton: Princeton University Press, 1995.

Proust, Marcel. "The Guermantes Way" Pt II, *Remembrance of Things Past.* Trans. C.K. Scott Moncrieff. London: Chatto and Windus, 1922.

R. L. L. "The Bible and Lord Byron." *The Imperial Magazine* 10, no. 116 (1828): 699–701.

"R.W". "*On the Modifications of Clouds, and on the Principles of Their Production, Suspension, and Destruction: Being the Substance of an Essay Read before the Askesian Society in the Session 1802*, 1803. By Luke Howard, Esq. 8vo. 2s.9d. Arch." *Monthly Review, or Literary Journal* 47 (1805): 432–433.

Raffles, Thomas Stamford. *The History of Java.* Vols. I & II, 2nd ed. London: John Murray, 1830.

Raffles, Sophia. *Memoir of the Life and Public Services of Sir Thomas Stamford Raffles* (1859) with introduction by John Bastin. Singapore: Oxford University Press, 1991.

Rampino, Michael R., Stephen Self, and Richard B. Stothers. "Volcanic Winters." *Annual Review of Earth and Planetary Sciences* 16 (1988): 73–99.

Reill, Peter Hans. *Vitalizing Nature in the Enlightenment.* Berkeley: University of California Press, 2005.

Reiman, Donald H., ed. *The Romantics Reviewed*, 9 vols. New York: Garland Publishing, 1972.

Reno, Seth T. "Romantic Clouds: Climate, Affect, Hyperobjects." In *Romantic Sustainability: Endurance and the Natural World, 1790–1830*, ed. Ben P. Robertson, 37–58. Lanham, MD: Lexington Books, 2016.

Ricoeur, Paul. *The Rule of Metaphor: Multi-Disciplinary Studies in the Creation of Meaning in Language* (1975). Trans. Robert Czerny with Kathleen McLaughlin and John Costello. London: Routledge and Kegan Paul, 1978.

Rigby, Kate. "Romanticism and Ecocriticism." In *The Oxford Handbook of Ecocriticism*, ed. Greg Garrard, 60–84. Oxford: Oxford University Press, 2014.

Robertson, Lisa Ann. "'Sensible' Slavery." *Prose Studies* 29, no. 2 (2007): 220–237.

Robertson, Ben P., ed. *Romantic Sustainability: Endurance and the Natural World.* Lanham, MD: Lexington Books, 2016.

Roe, Nicholas. *John Keats and the Culture of Dissent.* Oxford: Clarendon Press, 1997.

———. "John Keats's 'Green World': Politics, Nature and the Poems." In *The Challenge of Keats: Bicentenary Essays 1795–1995*, ed. Allan C. Christensen, Lilla Maria Crisafulli Jones, Giuseppe Galigani, and Anthony L. Johnson, 61–77. Amsterdam: Rodopi, 2000.

Rohrbach, Emily. *Modernity's Mist: British Romanticism and the Poetics of Anticipation.* New York: Fordham University Press, 2016.

Rousseau, Jean-Jacques. *Julie, or, the New Heloise: Letters of Two Lovers Who Live in a Small Town at the Foot of the Alps.* Ed. Philip Stewart and Jean Vaché, in *The Collected Writings of Rousseau*, ed. Roger D. Masters and Christopher Kelly, 13 vols., Vol. 6. Hanover: University Press of New England, 1997.

Ruskin, John. *Works*, 39 vols. Ed. E.T. Cook and Alexander Wedderburn. London: George Allen, 1903–1912.

Salisbury, William. [Letter]. *London Medical and Physical Journal* 35, no. 5 (May 1816a): 430.

———. [Letter]. *London Medical and Physical Journal* 35, no. 6 (June 1816b): 516–517.

———. [Letter]. *London Medical and Physical Journal* 36, no. 1 (July 1816c): 84–86.

———. [Letter]. *London Medical and Physical Journal* 36, no. 2 (August 1816d): 171–173.

———. [Letter]. *London Medical and Physical Journal* 36, no. 3 (September 1816e): 258–260.

———. [Letter]. *London Medical and Physical Journal* 36, no. 4 (October 1816f): 348–350.

Schama, Simon. *Landscape and Memory.* New York: A.A. Knopf: Distributed by Random House, 1995.

Schock, Peter A. *Romantic Satanism: Myth and the Historical Moment in Blake, Shelley and Byron.* Houndmills, Basingstoke, Hampshire; New York: Palgrave Macmillan, 2003.

Scott, Walter. "1. *Childe Harold's Pilgrimage*, Canto III. 8vo. 2. *The Prisoner of Chillon, a Dream; and Other Poems.* By Lord Byron. 8vo. John Murray: London." *The Quarterly Review* 16, no. 31 (1816): 172–208.

Scott, Heidi C.M. *Chaos and Cosmos: Literary Roots of Modern Ecology in the British Nineteenth Century.* University Park: Pennsylvania State University Press, 2014.

Secord, James. *Visions of Science: Books and Readers at the Dawn of the Victorian Age*. Chicago: University of Chicago Press, 2015.
Serres, Michel. "Science and the Humanities: The Case of Turner." *SubStance* 26, no. 2 (1997): 6–21.
Shakespeare, William. *The Tempest*. London: Methuen (Arden Shakespeare), 1976.
———. *The Norton Shakespeare*. Gen. Ed. S. Greenblatt. New York and London: Norton, 1997.
Sheats, P.D. *The Making of Wordsworth's Poetry, 1785–1798*. Cambridge, MA: Harvard University Press, 1973.
———. "'Tis Three Feet Long, and Two Feet Wide': Wordsworth's 'Thorn' and the Politics of Bathos." *The Wordsworth Circle* 22, no. 2 (1991): 92–100.
———. "Cultivating Margaret's Garden: Wordsworthian 'Nature' and the Quest for Historical 'Difference'." In *Placing and Displacing Romanticism*, ed. P.J. Kitson, 16–32. Hampshire: Ashgate, 2001.
Shelley, Percy Bysshe. "Journal at Geneva." In *The Complete Works of Percy Bysshe Shelley*, ed. Roger Ingpen and Walter Peck, Vol. VI. New York: Gordian Press, 1965.
———. *Shelley's Poetry and Prose: Authoritative Texts, Criticism*. Ed. Donald H. Reiman and Sharon B. Powers. New York: Norton, 1977.
———. *Selected Poetry and Prose*. London and New York: Routledge, 1991.
Shelley, Mary Wollstonecraft. "The Last Man." In *Novels and Selected Works of Mary Shelley*, ed. Jane Blumberg and Nora Crook, 8 vols., Vol. 4. London; Brookfield, VT: W. Pickering, 1996.
———. *Frankenstein, or the Modern Prometheus: The 1818 Text*. Ed. Marilyn Butler. Oxford: Oxford University Press, 1998.
Shelley, Percy Bysshe. *The Complete Poetry of Percy Bysshe Shelley*, 3 vols. Ed. Donald H. Reiman, Neil Fraistat, and Nora Crook. Baltimore: The Johns Hopkins University Press, 2003–2012.
Shelley, Mary Wollstonecraft. *Frankenstein*, or *The Modern Prometheus: The 1818 Version*, 2nd ed. Ed. D.L. Macdonald and Kathleen Scharf. Peterborough, ON: Broadview Press, 2005.
———. *The Annotated Frankenstein*. Ed. Ronald Levao and Susan J. Wolfson. Cambridge, MA: The Belknap Press of Harvard University Press, 2012.
———. *Frankenstein*. Ed. Maurice Hindle. London: Penguin, 2013.
Smith, Charlotte. *The Collected Letters of Charlotte Smith*. Ed. Judith Philips Stanton. Bloomington and Indiana: Indiana University Press, 2003.
———. *Charlotte Smith: The Major Works*. Ed. Claire Knowles and Ingrid Horrocks. Peterborough, ON: Broadview Press, 2017.
Smith, K.E. "'A Pile of Better Thoughts': Margaret, Silent Suffering and Silent Blessing." *Charles Lamb Bulletin* 145 (2009): 40–50.
St Clair, William. *The Godwins and the Shelleys: The Biography of a Family*. London: Faber & Faber, 1989.

Sutherland, John. *Is Heathcliff a Murderer?: Great Puzzles in Nineteenth-Century Literature*. Oxford: Oxford University Press, 1996.

———. *Who Betrays Elizabeth Bennet?: Further Puzzles in Classic Fiction*. Oxford: Oxford University Press, 1999.

Swann, K. "Suffering and Sensation in *The Ruined Cottage*." *PMLA* 106, no. 1 (January 1991): 83–95.

Tauchert, Ashley. *Romancing Jane Austen: Narrative, Realism, and the Possibility of a Happy Ending*. Houndmills: Palgrave Macmillan, 2005.

Terry, Judith. "Introduction." In *Journal of a West India Proprietor*, ed. M.G. Lewis, ix–xxxiv. Oxford: Oxford University Press, 1999.

Thompson, E.P. *The Making of the English Working Class*. New York: Vintage, 1966.

Todd, Janet. "Introduction." In *Persuasion*, 1818, ed. Jane Austen, xxi–lxxxv. Cambridge: Cambridge University Press, 2006.

Toohey, Elizabeth. "Emma and the Countryside: Weather and a Place for a Walk." *Persuasions* 21 (1999): 44–52.

Townshend, Dale. "Conjuration and Exorcism: Byron's Spectral Rhetoric." In *Byron's Ghosts: The Spectral, the Spiritual and the Supernatural*, ed. Gavin Hopps. Liverpool: Liverpool University Press, 2013.

Tremper, Ellen. "'Who Lived at Alfoxton?': Virginia Woolf and English Romanticism." London: Associated University Press, 1998.

Tuite, Clara. "*Sanditon*: Austen's Pre–Post Waterloo." *Textual Practice* 26, no. 4 (2012): 609–629.

Veale, Lucy, and Georgina H. Endfield. "Situating 1816, the 'Year Without Summer', in the UK." *The Geographical Journal* 182, no. 4 (December 2016): 318–330.

Vincent, Patrick. *La Suisse vue par les écrivains de langue anglaise*. Lausanne: Presses polytechniques et universitaires romandes, 2009.

White, R.S. "Keats and the Crisis of Medicine in 1815." *Keats-Shelley Review* 13 (1999): 58–75.

Whitehead, Scott. "1816: The Year Without a Summer in London." *Wanstead Meteo: Meteorology-Based Musings about East London and Beyond* (Blog). November 12, 2015. https://wansteadmeteo.wordpress.com/2015/11/12/1816-the-year-without-a-summer-in-london/.

Williams, Raymond. *Marxism and Literature*. Oxford: Oxford University Press, 1978.

———. *Problems in Materialism and Culture*. London: Verso, 1982.

Williams, J. "Salisbury Plain: Politics in Wordsworth's Poetry." *Literature and History* 9, no. 2 (1983): 164–193.

Wood, Gillen D'Arcy. "The Volcano Lover: Climate, Colonialism, and the Slave Trade in Raffles's History of Java (1817)." *Journal for Early Modern Cultural Studies* 8, no. 2 (Fall–Winter 2008): 33–55.

Wood, Michael. "Time and Her Aunt." In *A Companion to Jane Austen*, ed. Claudia L. Johnson and Clara Tuite, 195–205. Chichester: Blackwell, 2009.

Wood, Gillen D'Arcy. *Tambora: The Eruption That Changed the World*. Princeton: Princeton University Press, 2014.
Woolf, Virginia. *The Diary of Virginia Woolf, Volume Three: 1925–1930*. Ed. Anne Olivier Bell. San Diego: Harcourt Brace, 1981.
———. *The Common Reader*. Ed. Andrew McNeillie. London: Harcourt, 1984.
———. *The Essays of Virginia Woolf, III: 1919–1924* and *IV: 1925–1928*. Ed. Andrew McNeillie. London: Hogarth Press, 1988–1994.
———. *Orlando: A Biography*. London: Hogarth Press, 1992.
Wordsworth, William. *The Salisbury Plain Poems*. Ed. S. Gill. Ithaca, NY: Cornell University Press, 1975.
———. *The Ruined Cottage and The Pedlar*. Ed. J. Butler. Ithaca, NY: Cornell University Press, 1979.
———. *The Borderers*. Ed. R. Osborne. Ithaca, NY: Cornell University Press, 1982.
———. *Poems, in Two Volumes, and Other Poems*, 1800–1807. Ed. Jared Curtis. Ithaca, NY: Cornell University Press, 1983.
———. *Descriptive Sketches*. The Cornell Wordsworth, 21 vols. Vol. 8. Ed. Eric Birdsall and Paul M. Zall. Ithaca, NY: Cornell University Press, 1984a.
———. *The Major Works*. Ed. Stephen Gill. Oxford: Oxford University Press, 1984b.
———. *The Fourteen-Book Prelude*. Ed. W.J.B. Owen. Ithaca, NY: Cornell University Press, 1985.
———. *Lyrical Ballads, and Other Poems*. Ed. J. Butler and K. Green. Ithaca and London: Cornell University Press, 1992.
———. *The Fenwick Notes of William Wordsworth*. Ed. Jared Curtis. Tirril: Humanities-Ebooks, 2007.
Wordsworth, William, and Samuel Taylor Coleridge. *Lyrical Ballads*. London, 1798.
Wordsworth, William, and Dorothy Wordsworth. *The Letters of William and Dorothy Wordsworth: The Middle Years, Part II*, 2nd ed. Ed. Ernest De Selincourt. Rev. Mary Moorman and Alan G. Hill. Oxford: Clarendon Press, 1970.
Zerefos, C.S. et al. "Atmospheric Effects of Volcanic Eruptions as Seen by Famous Artists and Depicted in Their Paintings." *Atmospheric Chemistry and Physics* 7, no. 15 (2007): 4027–4042.

Index[1]

A

Abolition Act (1833), 109
Abolitionism, 93, 110n8
Abrams, M.H., 175
The Alps, 13
  in Lord Byron's letters, 126
  See also Swiss Alps
Anthropocene, 2, 15n1, 174, 189, 198, 199
Anthropocentric worldview, 6
Anthropogenic climate change, 2
Apocalypse
  in 'Darkness' (Byron), 163
  and language, 159
  in *The Last Man* (Shelley), 144, 146
Atmosphere
  and ambiguity, 174
  and communication, 187
  metaphoric uses, 174
Austen, Jane (1775–1817), 3, 88n12
  and aging, 82
  and aristocracy, 81
  and consciousness, 82, 87
  and contingency, 87
  Emma (1815), 76; meteorological analysis of, 76
  and free indirect discourse, 131, 134, 135
  and geology, 139n36
  global circulation of, 135
  intertextual connections with Lord Byron, 115
  meteorological analysis of, 76
  and the navy, 83
  non-linearity, 79
  Persuasion (1818), 9, 10, 77, 80–82, 84–87, 129, 130, 134, 135; citation of Byron in, 134; communication in, 129; drafting and completion of, 76,

[1] Note: Page numbers followed by 'n' refer to notes.

A. Collett, O. Murphy (eds.), *Romantic Climates*,
https://doi.org/10.1007/978-3-030-16241-2

Austen (*cont.*)
114, 132; gender relations in, 131; historicism in, 78; narratological innovations in, 79; representation of consciousness in, 80; representation of time in, 9, 12, 133; seasonal imagery in, 89n14; seasonal variation, 78
references to historical events, 76
representation of time, 77, 79, 80, 82, 83, 85–87, 88n12
response to 'Year Without a Summer', 79
Sanditon (comp. 1817), 87
and seasonal change, 81, 84
seasonal imagery, 80
and weather, 75, 88n1

B

Bate, Jonathan, 4, 8, 38, 39, 44, 61, 69, 170n17, 189n1
on significance of Mt. Tambora for 'Darkness' (Byron), 173
on 'To Autumn' (Keats), 69
Bees
influence on contemporary poetry, 70
'I stood tip-toe upon a little hill' (1817); representation of vision in, 64
and 1816's climate, 63
Poems (1817); nature as figurative presence in, 65
as symbol of climate change, 63
in 'To Autumn' (Keats), 69
Wordsworth on 1817 honey shortage, 63
Benjamin, Walter (1891–1940)
on aura, 181
on bourgeois interiors, 176
Birds
as victim of climate change, 72
Blackwood's Magazine, 115, 176
Blake, William (1757–1827)
The Book of Thel (1789), 25
Botany, 5
as medical discipline, 60
Brandes, Heinrich (1777–1834)
and pioneering weather map, 195
British Review, 120
Brougham, Henry (1778–1868), 115
Büchner, Georg (1813–1837), 162, 169n8
Burke, Edmund (1730–1797), 35, 53
Philosophical Enquiry into the Origin of Our Ideas of the Sublime and Beautiful (1757), 21
Reflections on the Revolution in France (1790), 183
on the sublime, 21
Burnet, Thomas (1635–1715), 147
on mountains, 147
Telluris Theoria Sacra (1681–89), 147
Burns, Robert (1759–1796), 43
Butler, Marilyn, 136
Byron, Lord (1788–1824), 108, 126, 141–143, 161, 163, 164, 169n12
Alpine Journal (1816), 117, 123, 142, 149, 150
Augustus Darvell (1819), 114, 124
Cain (1821), 123, 138n25; as prequel to Manfred (1817), 137n2
The Corsair (1812), 134
'Darkness' (1816), xi, 4, 13, 14, 113, 124, 159, 161, 164, 165, 167, 173; biblical tropes in, 163; as climate change poem, 191, 195; commentary on atmospheric phenomena in, 162; divergence from religious tradition in, 163; early biblical

interpretations of, 163; solipsism in, 167; spectatorship in, 165; use of Biblical tropes in, 164
death, 146
Don Juan (1819–1824), 113, 127–129; on temporality of love in, 128
A Dramatic Poem (1817), 124
exile in Switzerland, 114
'English Bards and Scotch Reviewers' (1809), 92, 114
and exile, 122
and exorcism, 138n24
and the failure of language, 167
The Giaour (1813), 132
influence on geology, 118
intertextual connections with Jane Austen, 115
Manfred, A Dramatic Poem (1817), 12, 114, 117–122, 126, 127, 133, 149–151, 166; contemporary reviews of, 120; demonic imagery in, 150; incest in, 119; reviews of, 118; Switzerland in, 151
and Napoleon, 132, 133
observes Swiss Alps, 143
'Ossian's Address to the Sun' (1805), 165
parody of Walter Scott, 109n6
scandalous rumours surrounding, 121
separation from Lady Byron, 127
'Some Observations upon an Article in Blackwood's Edinburgh Magazine' (comp. 1820), 115
use of atmospheric phenomena as metaphor for apocalypse, 163
'A Version of Ossian's Address to the Sun' (1806), 165

C
Caribbean
hurricane activity, 103
Catastrophism, 93, 115, 127
in Lord Byron, 113, 118, 123
Chambers, Robert (1802–1871), 21
Chandler, James, 40
reads Wordsworth as Burkean conservative, 35
Chaos theory
in Austen's *Persuasion* (1818), 87
and literature, 79, 80
Cholera
origins of epidemic, 193
Clarke, Gillian, 70
Climate, 1, 3, 5, 15
and early readings of *Persuasion* (1818), 81
and effect on aging in Austen's *Persuasion* (1818), 84
in 1816, 192
gendering in Austen's *Persuasion* (1818), 83
in novels of Jane Austen, 77
Climate change, viii, 8, 76, 83, 175
in contemporary poetry, 70
future consequences, 196, 197
in 1816, 18
in John Keats's work, 61
modern crisis, 175
in *Orlando* (Woolf), 178
parallels between 1816 and present day, 193, 194
poetry of, 72
and political paralysis, 197
in present day, 196
and quasi-chaotic causal physics, 198
rapidity, x
and Romanticism, 189
and Romantic scholarship, 61

Clouds, 7, 24, 177
Howard's categorisation of, 23
Howard's conception of, 26
in Keats, 25
linguistic instability, 24
in Lord Byron, 123
Coleridge, Samuel Taylor (1772–1834), 36, 179
'This Lime-Tree Bower my Prison' (1800), 180
'The Rime of the Ancient Mariner' (1798), 70
Cowper, William (1731–1800)
The Task (1785), 173
Coxe, William (1747–1828)
Sketches of the Natural, Civil, and Political State of Swisserland (1779), 148
Cuvier, Georges (1769–1832), 20

D
Davy, Humphry (1778–1829), 124
proof-reads *Lyrical Ballads* (Wordsworth and Coleridge), 174
Deep time, 119, 134
in geological thought and Romantic poetry, 118
Deleuze, Gilles, 184
Duffy, Carol Ann (1955–), 70

E
Eagleton, Terry, 39
Ecocriticism, 16n3, 38, 61, 79, 198, 199
as response to ecological catastrophe, 162
Ecological science
and 'teleconnection', 116
Ecology
in Austen's *Persuasion* (1818), 79, 83, 87
Edinburgh Review, 119
Enlightenment, 7, 15
and mechanisitic view of nature, 20
Environment, viii
and embeddedness, ix
as web of experiences, 6
Erskine, Thomas (1750–1823), 50, 53
Exile
in Lord Byron's writing, 116

F
Famine, 196
Forgetting
in *Persuasion* (Austen) and *Manfred* (Byron), 133
Franklin, Benjamin (1706–1790), 162
Franklin, Caroline, 81
Free indirect discourse
and time, 140n38
Frend, William (1757–1841), 50, 53

G
Galperin, William, 86
Gender relations
in Austen's *Persuasion* (1818), 82
in *Orlando* (Woolf), 181
Gentleman's Magazine, 119
Geology, 5, 19, 119
catastrophism, 11, 12, 20, 117
as challenge to Christianity, 138n18
gradualism, 11, 19
Gerrald, Joseph (1763–1796), 45, 46, 53
Global cooling, 17
Godwin, William (1756–1836), 13
on *Paradise Lost*, 153
Goethe, Johann (1749–1832), 26

The Sorrows of Werther (1774), 132
Gray, Thomas (1716–1771)
Elegy in a Country Church-Yard (1751), 43
Guattari, Félix, 184

H

Harris, Jocelyn
argues for significance of Napoleon as absence in *Persuasion*, 132
Hegel, G.W.F. (1770–1831)
Phenomenology of Spirit (1807), 166
Heidegger, Martin (1888–1976), 170n16
and metaphysics, 167
Historical change, 9
Historicism, 10
in Austen's *Persuasion* (1818), 78
Hobhouse, John (1786–1869), 142, 149, 150
Holland, Lady Elizabeth (1771–1845), 97
Holland, Lord Henry (1773–1840), 97
Honey, *see* Bees
Hopkins, Gerard Manley, 70
Howard, Luke (1772–1864), 7, 17, 18, 21, 25
The Climate of London (1818–1820), 18–20, 22, 23, 27, 28, 29n1, 29n3, 76
mechanistic view of climate, 29
mechanistic view of nature, 20
objects to agential nature, 20
On the Modifications of Clouds (1803), 20, 23, 24, 27
religious idea of nature, 21
Humboldt, Alexander von (1769–1859), 20
Hunt, Leigh (1784–1859), 64
Hutton, James (1726–1797)
Theory of the Earth (1795), 31n38, 117
Huxley, Aldous (1894–1963), 8, 40
and imperialist tropes, 35
'Wordsworth in the Tropics' (1929), 34–36, 54

I

Imperialism
and Matthew Lewis, 93
Industrial Revolution, 6
Interdisciplinary thought, 5, 6, 12
Interdiscursivity
as urgent project, xi
Intertextuality
among Romantic writers, 116
in *Orlando* (Woolf), 181

J

Johnson, Claudia
critiques perception of Austen's miniaturism, 135

K

Kafka, Franz (1883–1924)
'Wish to become an Indian' (1913), 184
Kant, Immanuel (1724–1804)
'Analytic of the Sublime' (1790), 166
Keats, John (1795–1821), 8, 179
begins medical studies, 60
and botany, 66
'Calidore' (1817), 180
on clouds, 25
connection between poetry and climate, 61

Keats (*cont.*)
and effects of 1816's climate, 69
influence on contemporary poetry, 70
'I stood tip-toe upon a little hill' (1817), 61, 62, 64
medical studies, 60
observational techniques, 64
'Ode to a Nightingale' (1819), 68
'Ode to Psyche' (1819), 67, 68, 72
Poems (1817), 63–65
response to climate of 1819, 66
'To Autumn' (1819), 68, 69, 71
'To Autumn' (1820), 25
'To Charles Cowden Clarke' (1817), 65

L

Lake District
in William Wordsworth's poetry, 7
Lake District Search and Mountain Rescue Association, 34, 55, 55n1, 58n67
Language
and experience, 14
and nature in *Orlando* (Woolf), 182
Lavoisier, Antoine (1743–1794), 20
Lewis, Matthew (1775–1818), 91–93, 101
abolitionism, 97
ambiguous attitude towards slavery, 107
anticipates *Heart of Darkness* (Conrad), 107
considers emancipation impossible, 97
dies of yellow fever, 107
embarks on first Jamaican voyage, 98
historical disturbances in *The Journal of a West-India Proprietor* (1834), 109
influence of Edmund Burke on, 101
inherits sugar plantations, 96
Journal of a West-India Proprietor (1834), 10, 92, 93, 96–101, 103–108
Journal of a West-India Proprietor as document of British imperial project, 94
as liberal slave-owner, 93, 96, 106
makes return voyage to England, 106
The Monk (1796), 93
racist assumptions, 102
representation of animals, 99
representation of Jamaica, 105
views on abolition and emancipation, 93
warns Lord Holland, 97
Liverpool, Lord (1770–1828)
and political repression, 194
London
historical weather patterns, 29n2
in *Orlando* (Woolf), 175
London Corresponding Society, 45
Longley, Michael, 70
Lucretius
De Rerum Natura, 113
Lyell, Charles (1797–1875)
Principles of Geology (1830–1833), 19, 117, 118
Lyme Regis, 85

M

Macdonald, D.L., 93
Macpherson, James (1736–1796), 165
Materialism
in Wordsworth's poetry, 40
McGann, Jerome, 166
Measurement, 7
Metaphor, 168n5, 170n13
as 'species of catachresis', 160

Meteorology, 5, 19
origins, 174
Milton, John (1608–1674)
on mountains, 147
Paradise Lost (1667), 147, 151
Paradise Regained (1671), 147
Romantic readings of *Paradise Lost*, 152
Mont Blanc, 21, 125, 126, 154, 156n14
in Mary Shelley's *Frankenstein* (1818), 154
seen as cursed, 148
Moore, Thomas (1779–1852), 116
uses *Alpine Journal* to explicate *Manfred*, 117
More, Hannah (1745–1833), 45, 52
Village Politics (1792), 45, 46
Morton, Timothy, 3, 16n4, 40, 56n22, 56n23, 123, 139n26, 166
Mountains, 142
biblical interpretations, 147
as observed in *The Last Man* (Shelley), 145
representation to late seventeenth century, 147
and the sublime, 19

N
Napoleon, 161
Napoleonic wars, 77
Nature
and authenticity, 40
and language, 186
in Lord Byron, 169n10
as metaphor, 3
and presentism, 40
problems of representation, 187
relationship with art, 9
and technological processes, 174
Nicolson, Marjorie Hope
Mountain Gloom and Mountain Glory (1959), 147

O
Oak tree
as English symbol, 183
varying meanings in *Orlando* (Woolf), 188
Objectivity
eighteenth century notion of, 26

P
Paley, William (1743–1805), 22, 45, 52
Reasons for Contentment; Addressed to the Labouring Part of the British Public (1792), 45
Peel, Sir Robert (1788–1850), 195
Polidori, John (1795–1821), 108, 125
Ernestus Berchtold (1819), 149
The Vampyre (1819), 5, 114
Postcolonialism
conceptual limitations in Anthropocene, 197

R
Raffles, Lady Sophia (1786–1858), 96
Memoir of the Life and Public Services of Sir Thomas Stamford Raffles (1859), 94, 110n11
Raffles, Thomas Stamford (1781–1826), 94, 95, 110n9
abolitionist policies, 95
Robertson, Lisa Ann, 97
Romantic
readings of Milton, 152
Romanticism, 116
in the Anthropocene, 198, 199

Romanticism (*cont.*)
critical mythologies of, 8, 36, 54
and geology, 117
language in, 187
and nature, 186
Romantic literature
and climate change, 173
Romantic movement
representation of natural world as sublime, 19
Romantic period
and climactic vocabulary, 174
and fossil-based energy, x
Romantics
representations of Satan, 153
Rousseau, Jean Jacques (1712–1778), 142, 155
Julie, ou la nouvelle Héloïse (1761), 145
Ruskin, John (1819–1900), 15, 169n10
Modern Painters (1836–1860), 176
on the pathetic fallacy, 177, 178
on poetry, 177
on Romantic artworks, 177

S

Satan
radical representations in Age of Revolutions, 152
Romantic images of, 155
Scarcity
in 1816–17, 196
Schama, Simon
Landscape and Memory (1995), 143
Scheuchzer, Johann Jakob (1672–1733), 148
Scott, Walter (1771–1832), 161, 179
Anne of Geierstein (1829), 155
on Lord Byron, 159
Shakespeare, William (1564–1616)
King Lear (1606), 41, 42
Macbeth (1606), 106
The Tempest (1611), 10, 101, 103, 108
Shelley, Mary (1797–1851), 115, 149, 192, 194
apocalyptic themes, 144
Frankenstein (1818), xiii, 5, 12, 114, 144, 149, 151–153, 173; first theatrical adaptation of (1823), 153; parallels with *Paradise Lost*, 151; satanic relationships in, 152; Switzerland in, 154; vitalism in, 20
The Last Man (1826), 12, 13, 144, 145; disillusionment in, 146
and loss, 146
and Mary Wollstonecraft, 131
and nostalgia, 146
representation of 1816, 146
on summer of 1816, 114
use of alpine landscapes, 153
Shelley, Percy Bysshe (1792–1822), xiii, 13, 108, 115, 162, 192
'The Cloud' (1820), 25
death, 146
'Mont Blanc' (1817), x, 21
Prometheus Unbound (1820), 3
St. Irvyne (1811), 149
Slave Felony Act (1811), 95
Slavery Abolition Act (1834), 10
Smith, Charlotte (1749–1806)
'Sonnet 43' ('The unhappy exile, whom his fates confine...') (1784), 25
Society of West India Planters and Merchants, 97
Southey, Robert (1774–1843), 116, 123
Souveniring
as mode of remembering, 126
Spenser, Edmund (1552–1599)
The Faerie Queene (1590), 42
Subjectivity, 14

Suffering, 8, 10
Swiss Alps, 148, 149, 155
  changing ideas of, 155
  as demonic, 152
  as dystopia, 147
  as emblem of the sublime, 141
  perceived as dystopia, 142
  as seen by Lord Byron, 143
  seen by Romantics as sublime, 149
Switzerland, 13, 146
  dualities, 156
  as dystopia, 150, 155
  in 1816, 141, 192
  in *The Last Man* (Shelley), 144, 145
  read as topological embodiment, 155
  republican heritage, 154
  in Romantic period, 155
  as utopia, 141, 143, 145, 146, 155
  visited by Lord Byron and Mary Shelley in 1816, 142
Sympathy
  as creative response to events of 1816, 194, 195

T

Tambora, Mt., 1–4, 41, 59, 61
  climatological impacts, 162
  consequences of 1815 eruption, vii, 195
  eruption, 62, 76, 91, 93, 160, 167; consequences of, vii, 30n6, 193; effects of, 117, 167; effects of, in Atlantic, 102
  global effects of eruption, 167, 192, 193
  impact of eruption on Indonesian archipelago, 95
  impact on Indonesian archipelago, 94
  philosophical implications, 136
Tauchert, Ashley, 81
Teleconnections, 196–199
  and new global historiography, 198
  and weather anomalies, 137n6
Temporal discontinuity
  in Romantic artwork, 178
Tennyson, Alfred (1809–1892), 179
Textual influence, 140n44
  and displacement of chronology, 116
Time
  in Austen's *Persuasion* (1818), 80
Todd, Janet, 132
Travel
  in Austen's *Persuasion* (1818), 85
Turner, J.M.W. (1775–1851), 173
  'The Slave Ship' (1840), 105

V

Villa Diodati, xiii, 4, 91, 108, 114–117, 142, 160, 163, 191, 195
  Diodati circle, 4, 10, 12, 13, 92, 144, 146
Vitalism, 20, 21
  in *Frankenstein* (Shelley), 30n15
Volcanic activity, 168n7
Volcanic events
  and atmospheric change, 162
Volney, des comte
  Ruins of Empires (1793), 132

W

Waterloo, 77, 161
  geopolitical and cultural impact, 77
Weather, 14
  in *Orlando* (Woolf), 175
Wilberforce, William (1759–1833), 97, 108
Williams, Raymond, 178, 182, 183, 187
  on changing ideas of nature, 174

Wollstonecraft, Mary (1759–1797), 131
Wood, Gillen D'Arcy, vii, 4, 17, 95, 96, 116, 117, 160, 161
on Luke Howard, 23
Woolf, Virginia (1882–1941)
on atmosphere in English literature, 187
blurring of distinctions, 179
blurring of historical distinctions, 188
depiction of climate change, 179
and Enlightenment imagery, 175
and 'haecceity', 184
and historical context, 15
on nineteenth-century prose style, 176
on the nineteenth-century subject, 176
on oak trees, 182
Orlando (1928), 14, 176, 184–187, 189; climate change in, 176; environmental themes in, 181; first edition of, 186; gender instability in, 181; intertextuality in, 185, 189; literary composition in, 182; nature in, 179, 188; oak trees in, 182, 185; parodic elements in, 188; weather and culture in, 178
and the prospect poem, 183
representation of historical experience, 188
representation of time, 181
resists stable signifiers, 188
reworkings of Romantic poetry, 175
and Romanticism, 188
use of satire, 179
on writing and nature, 186
Wordsworth, William (1770–1850), 8, 148, 155, 175, 179
The Borderers (1797), 53
depiction of socio-economic reality, 44
Descriptive Sketches (1793), 148, 151
on Edmund Burke, 182
on 1817 honey shortage, 63
and generic subversion, 43
and history, 41
and homelessness, 44
idea of nature, 34; read as conservative, 37
landscapes and weather in 1790s poetry, 41
Lyrical Ballads (1798), 43, 174
radical idea of landscape, 42, 47
reflects on childhood, 185
representation of poverty, 47
'The Ruined Cottage' (1797), 46, 49, 53; conjunction of nature and history in, 49, 52
Salisbury Plain (1793–94), 41–44, 46, 47, 51, 56n26, 56n29, 57n32
and Spenserian verse form, 42
and suffering, 42
on Switzerland, 148
'The White Doe of Rylstone' (1815), 180
'Tintern Abbey' (1798), 8, 35, 37, 54, 180, 181
'To Toussaint L'Ouverture' (1802), 54
'The world is too much with us' (1802), xi, 6, 16n9

Y

Yellow fever, 111n35

Z

Zong, massacre of the, 105

The manufacturer's authorised representative in the EU is Springer Nature Customer Service Centre GmbH, Europaplatz 3, 69115 Heidelberg, Germany. If you have any concerns regarding our products, please contact ProductSafety@springernature.com

Printed and bound by CPI Group (UK) Ltd, Croydon, CR0 4YY

05/10/2026

02228125-0003